Creating and Leading
High-Performance Organizations

CREATING AND LEADING
HIGH-PERFORMANCE
ORGANIZATIONS

Cecil V. "Buddy" Martinette Jr.

BOOKS & VIDEOS

Copyright © 2023 by
Fire Engineering Books & Videos
110 S. Hartford Ave., Suite 200
Tulsa, Oklahoma 74120 USA

800.752.9764
+1.918.831.9421
info@fireengineeringbooks.com
www.FireEngineeringBooks.com

Executive Vice President: Eric Schlett
Vice President, Group Publishing: Amanda Champion
Acquisitions: David Rhodes, Diane Rothschild
Director, eLearning and Books: Starlet Franz
Production Manager: Tony Quinn
Developmental Editor: Chris Barton
Operations Manager: Holly Fournier
Cover Designer: Brandon Ash
Book Designer: Robert Kern, TIPS Publishing Services, Carrboro, NC

Library of Congress Cataloging-in-Publication Data

Names: Martinette, C. V., Jr., author.
Title: Creating and leading high-performance organizations / Cecil V. "Buddy" Martinette.
Description: Tulsa, Oklahoma : Fire Engineering Books & Videos, 2023. | Includes index.
Identifiers: LCCN 2023029132 | ISBN 9781593705855 (paperback) | ISBN 9781593705855 (epub)
Subjects: LCSH: Leadership. | Martinette, C. V., Jr. | Fire departments--Management--Case studies.
Classification: LCC HD57.7 .M37 2023 | DDC 658.4/092--dc23/eng/20230707
LC record available at https://lccn.loc.gov/2023029132

Printed in the United States of America

1 2 3 4 5   26 25 24 23

Although my failures have been many there is always one person by my side. A person who loves me unconditionally and I can always count on for emotional, physical, and spiritual support. I dedicate this book to the one person who has never let me down, my wife and life partner, Sarah Martinette.

Contents

Part I
Introduction: The High-Performance Way

Part II
Personal Leadership: Getting the Ingredients Right

Part III
Organizational Leadership: Creating the Right Environment

Part IV
Connecting Leadership: Making It All Work

Foreword

Chief Buddy Martinette was appointed fire chief for the Wilmington Fire Department, becoming the only fire chief hired from the outside in the nearly 125-year history of the department. Expectations were high for our new leader, as our department had just undergone a comprehensive assessment by a consulting group and their report outlined many pages of recommended changes. The department was eager for change, and this report would become the new chief's job description.

When Chief Martinette arrived in September of 2008, I was a captain assigned to the Training Division. This meant that I would be a member of his staff and, as such, would play an important role in implementing many of the chief's new programs and policies. I, too, was eager for change and had recently transferred from the Operations Division after 17 years, in hopes of helping the organization become more progressive. Never in a million years would I have imagined the transformational changes that lay ahead for our organization. Nor could I have imagined the profound impact working for Chief Martinette over the next 13 years would have on my personal and professional growth as a leader.

In 2011, I was in my first year of the National Fire Academy's Executive Fire Officer (EFO) Program. The first year's class was titled *Executive Development*. The focus of this course is to develop and strengthen chief fire officer leadership skills in the areas of critical thinking, managing change, high performance teams, valuing diversity, and adaptive leadership. By the end of the first week, it became very apparent to me that the chief I worked for was the very leader the EFO program sought to create. It came as no surprise when I noticed his picture on the EFO "Wall of Fame" for Outstanding Research Award winners. The work that Chief Martinette was doing in our department modeled the leadership lessons I was taught throughout my 4 years at the academy.

Chief Martinette is a transformational leader. His leadership transformed the culture of the Wilmington Fire Department from a rule-driven, very authoritarian organization to one based on core values, empowerment, and trust. That cultural change is his legacy, a legacy earned from many years of adaptive leadership, patterned behavior over time, vision, and the ability to effectively manage organizational change.

Today, the Wilmington Fire Department is a twice accredited and Insurance Services Office Class 1 Fire Department, placing us among the very best fire departments in the country.

As Chief Martinette's successor, I have very big shoes to fill. However, working alongside Chief Martinette for more than a dozen years has given me the knowledge, skills, experiences, and confidence necessary to lead our organization forward. I will be forever grateful for his leadership, coaching, mentoring, and most of all his friendship.

It is my sincere hope that the information in this book will not only help you become a better leader but inspire you to create an organizational culture based on core values, empowerment, and trust.

Respectfully,
Jon S. Mason
Fire Chief
Wilmington, NC, Fire Department

Preface:
Before We Get Started

We develop into the people we are because
of the experiences we live, and, in effect, we
become the sum of our experiences.

Unlike many leadership books, I won't profess that this one is the end-all, be-all. This book will not by itself make you a great leader. You see, to be a good leader is as much about the timing as it is about the actual things you do. I learned this lesson early on and just stuck with it because being unsuccessful was never an option. Inscribed on the back of my desktop nameplate is a quote by Winston Churchill: "Never, Never, Never Quit." A trait, I might add, that does not endear me to everyone who works with me or for me.

The other interesting aspect of leadership is that it evolves in much the same way as you grow and gain the benefit of experience and hindsight. I say this because as a lifelong learner, I now confess to questioning principles and practices that up until a few years ago I was certain were the best way to lead. As I gain experience and maturity, I get more perspective and view the world differently. Hurrah for experience and hindsight, huh?

This book is written so the many principles presented about fire service leadership can be applied to every kind of organization, business, or team. The thoughts and opinions herein are based on real-life experiences, and I am very proud to share them with you. If you would prefer only run-of-the-mill bullet points that constitute many of our profession's leadership books, then this book may not be the best option for you.

As you start reading, you should know that I have made mistakes in both my personal and professional life, but I have learned from them. In full disclosure, you should also know that even though I have written this book, I do not know everything about leadership. I am, and continue to be, far from a perfect leader. I do, however, think I have experienced enough about leadership to help you lay a framework by which you could potentially see a successful outcome from your efforts.

I am happy to present this book as a measure of gratitude to the fire service family that has cared for me and mentored me all these years. This book is truly shared out of love. You'll find in these pages stories and examples to help you build on your talents to lead people and inspire them. I hope that after reading these words you will be a better leader and therefore a benefit to others.

To be an effective leader, you first need to develop credibility, which is established by a willingness to open yourself up to others. That includes personal information squaring us up as human beings establishing a partnership. The information presented here at the beginning of our journey has the sole purpose of developing a relationship with you, the reader.

Part I

Introduction:
The High-Performance Way

1

The Story

The culture of an organization is established based on the collective experiences of its members in the same way as our life experiences and interactions with others define who we are and how we view the world.

Key Points

- Average people can do good things.
- When life stacks up against you, find a positive way forward.
- If you are willing to be helped, people will help you.
- Self-worth is born of being selfless.
- No leader is created entirely of their own effort, but rather by others who see value and invest in them.
- Hard work will always overcome a challenge.
- God has control, and we should allow him to control.

The Beginning

Life is good to you if you have the inner spirit and competitive drive to stick with your commitments, a spirit telling you to work harder, study longer, and try again when your aspirations fall short of a successful outcome. The success I refer to lies in the spirit of a person not afraid to fail because stopping at failure is never an option.

Growing Up

I am the product of a broken home. My parents separated soon after my younger brother, Brian, was born. My mom battled bipolar disorder, and my childhood was filled with daily chores that included taking care of my infant brother while my mom and a guy-of-the-week were at the local pub.

The effect of this environment resulted in my lack of self-confidence and any sense of responsibility or personal accountability. As a preteen, I had no drive or appreciation for life and no personal value system (fig. 1–1). Depression followed, with feelings of worthlessness and self-pity. A psychologist, a psychiatrist, and hospital personnel tried to figure out what was wrong with me. It turns out that time spent at the hospital may have saved my life, and it led to my conviction that God sends kids to places to protect them from their environment.

I had a set of grandparents who cared very much for me. I also had a father who—despite his son's head being filled with ideas that he was a terrible man—never gave up on the hope that he would have his day of redemption. I can never recall my father saying a bad word about my mother despite her ill will toward him. That was a powerful lesson, and my dad to this day continues to be an honorable Christian man.

My mom, who has since passed away, was not an evil person. She was sick and lived with a terrible illness that I'm sorry she had to face daily. I have come to understand that many of my experiences regarding my mom's mental illness have helped me as a leader in my adult life.

FIGURE 1–1. The author as a wayward teenager

I returned to my dad and stepmother, Marie, back in Virginia a tattered, confused, long-haired preteen with no manners or social skills. Marie acted like a drill sergeant set out to break me of my old habits. I thank her and my dad for sticking by me before I turned into a law-breaking, drug-taking, poorly behaved underachiever.

Within the next few years, I slowly regained a sense of purpose about myself and made it to high school.

How Community Service Changed Me

Without any clear direction or a social safety net, I repeatedly got in trouble and waited for the next bad thing to happen, spending my fair share of time in the office of the assistant principal (Mr. Shackleford) at Princess Anne High School in Virginia Beach, Virginia (fig. 1–2).

What could have caused a wayward young man like me to turn his life around? Did he witness a fiery car crash and pull somebody from the flaming wreckage? Was it a nearby building fire where he kicked the door in and save the many workers inside? I bet he noticed a fellow student choking and jumped to the rescue. No, it was much more dramatic than that. On that day, in front of that majestic brick school, I unceremoniously tossed an ice cream wrapper on the ground and

FIGURE 1–2. The author's alma mater wishes him and his peers luck for a school event.

started walking back to school, and I truly believe it was a higher power that changed my life with a teachable moment.

Mr. Shackleford's response that day would change everything about my life. He told me that I would be required to do something called community service. *What the heck is community service and how am I qualified to do it?* I wondered. "You are kidding, right? Just suspend me and let the both of us get on with our lives," I quipped. Mr. Shackleford responded, "You can pick up trash or feed the homeless if you want, but before you can participate in any more sports at this school you will complete 8 hours of community service." I thought this punishment was the most unfair of all for a littering violation.

A New Order: Shiny Red Fire Trucks and Cleanliness

On my way home from school that day, I walked by a fire station and noticed freshly washed fire trucks sitting on the apron, newly shined and clean (fig. 1–3). To this day, I don't think there is anything more beautiful than a shined-up red

FIGURE 1–3. The author's first view of new, shiny fire trucks outside Thalia Volunteer Fire Department

fire truck. There is just something that stirs a firefighter's soul when we see a fire truck with its polished chrome and gold leaf contrasting against a deep red background. All I know is that something that day pulled me toward that fire station and the open bays that sat behind those trucks.

I walked inside the bays, and the first thing I noticed was that the floors were painted gray and they shined in the same manner as the trucks. *What in the world, do the firefighters have a maid or cleaning crew for this place*? At this point, if you are a firefighter, you are laughing because you can picture me with a brush in my hand having cleaned thousands of toilets over the course of my career. If it just so happens you are not a firefighter, you should know that, yes, we do our own cleaning and, yes, that does include the toilets.

Man, this place is so clean. I mean really, who cleans stuff like this? I thought. *One more door to check in the back of the building and then I am out of here. But wait, should I open the closed door? What if they are in a meeting or something?*

Gathering courage, I pressed forward and cracked the door open gently. Single beds with red covers were neatly arranged in a long, spacious room. Every cover was folded the same way at the foot of the beds, each one creased perfectly and not a wrinkle in any of them. In between each bed was a nightstand that held a single lamp. The only other items in this very neatly arranged room were a small round table and three chairs. On that day, those chairs were occupied by the next set of individuals that God would bring into my life.

Charlie Carson, Skip Brehm, and Billy Mills sat at the table in that bunkroom reading *Playboy* magazines. I thought, at age 15, what a great gig these guys had going.

After a brief round of "What can we help you with, young man" conversation, I explained my situation at school and asked if I could sweep the floor or wash the trucks to complete my community service. "Nothing like that around here," one of them said. "But you can be a volunteer if you want. How old are you?" I told him I was 15, to which he countered with, "You need to be *16* to volunteer. How old are you?" "15," I replied again. He countered again with, "You need to be *16* to volunteer." We continued this Abbott and Costello script until I realized I was 16.

Those men, and many others after, took me in, mentored me, coached me, and in many cases fathered me through my adolescent years. Through their actions and behavior, they helped me understand that life and all its wonderful blessings are achieved through the giving of oneself to others.

For the next 3 years, I spent much of my time living and working in that fire station. I learned from people of all walks of life who all cared about the same thing, which was helping those in need. As I reflect on my years as a volunteer, and the subsequent years I have spent working with volunteers, I am always amazed by the self-sacrifice and humility displayed by those paying that sacrifice (fig. 1–4).

FIGURE 1–4. From left are Buddy Martinette, Frank Hendricks, Billy Mills, Tal Luton, and Ernie McGanty, officers of the 1978 Thalia Volunteer Fire Department

I've had a lifetime to think about service and leadership, starting as a troubled young man from a broken home. As I stood at that crossroad, it was my chance to either continue to be that troubled young man from a broken home doing things that would affect my life in a negative way and forever keep me from being successful or to choose a different road.

I decided to take the high road, and I have God—by way of Mr. Shackleford, Charlie, Skip, and Billy and those shiny red trucks—to thank. For the first time in my life, I found myself surrounded by people, other than my family, who cared more about me than I cared for myself. That day, with an ice cream wrapper, an assistant principal who refused to suspend me again, and an open fire station, would turn out to be one of the most important days of my life.

The First of Many Calls

I still remember my first call: Dispatch to Station 7, brush fire 3700 Block of Virginia Beach Boulevard. I ran straight over to the hook on the wall that held my gear. Canvas bunker coat, three-quarter boots, a plastic MSA helmet, and a pair of fireball gloves. For those of you too young to remember fireballs, they were red rubber gloves with no thermal protection. Looking back on it, I wonder what they hell we were thinking wearing combustible gear, no pants, and a helmet and gloves that would melt while you were wearing them and fighting fire.

Unit 704 was our brush truck. It was a medium duty pickup truck with 100 gallons of water and a top-mounted pump. It also had 150 ft. of rubber hose we called the booster line. My mission was to get to that truck.

I fell over trying to get my boots on, mostly because I was trying to get my coat on at the same time. I had prepared to get dressed as fast as I could at and at all costs, as I didn't want to be the rookie who got left at the station trying to get his gear on the right way. It was not a graceful fall; it was more of a splat!

When I finally stood up, I'm sure that I looked like that dude Arte on *Laugh-In* (fig. 1–5). My coat was almost down to my ankles, my boots fit like clown shoes, and I could just barely see out from under the brim of a helmet that was surely made for some gigantic man's head. I wasn't just a mess—I was a chocolate mess!

The passenger pointed toward the back of the truck. "In the back?" I asked. "That's right, in the back." And so, it was to be that my first lights-and-sirens real-life fire call was a long, very chilly ride in the back of a pickup truck along with the portable Indian Cans, rakes, and shovels.

FIGURE 1–5. The author felt like he was wearing clown shoes in his first firefighter outfit.

The red lights were on, and the siren wailed. People pulled their cars over to the side of the road so we could get to our emergency call as fast as possible. I was nervous as hell because not only was I on my first fire call, I was trying not to get thrown out of the back of the truck while also trying to keep that brush rake out of my butt.

I was participating in something worthwhile. For the first time in my life, I was doing something that I could be proud of and I was helping someone other than myself.

That call ended up being no big deal. Someone had stopped and put the fire out just as we were arriving. The thing I do remember most about that call was that I wanted someone to see me. I didn't care who it was, I just wanted someone to see me riding in the back of that truck. I so wished it could have been my grandfather.

There would be many more calls to come. My days after school and the nights that followed would be spent waiting for and responding to fires, accidents, medical calls, and just about every other type of calamity you could ever imagine. As a very young man, I saw dead people, people still alive who were crushed between cars, people standing in the front yards of their homes with skin melted away from their bodies, and others who could not escape their flaming homes and were burned beyond recognition. This is not the kind of stuff for the faint of heart, to say the least.

My First Blue Shirt

When I turned 19, I had the opportunity to become a paid firefighter. I was a dedicated volunteer and had almost all my certifications, and it was clear that the once infrequent call volume had given way to a very busy fire company. The old-timers had a hard time with the transition from volunteer to a combination system.

Because of the relationship my company had with the career system, I wasn't sure I would get picked up. I was leery of the negative associations among myself, the volunteer company, and the city fire department. The folks who came from our volunteer station were labeled as troublemakers, and there were many kids just like me who all had the same dream of being a paid firefighter.

Chief Diezel told us the first day on the job that all of us got there based on our own merits. Personally, I think he was being kind because many of us, including myself, had advocates who were connected to the political machine of the time. None of us in Virginia Beach Rookie School #13 would ever come to find out if we earned our way to that spot or not. As I recall, when the dust settled, none of us really cared; we were all happy to be getting a check for something as fun and rewarding as firefighting (fig. 1–6).

At the end of that first day, we were issued our uniforms and sent home with instructions with where we were to report. As quickly as possible, I took that light blue shirt out of its plastic wrap and placed Badge #56 in the pre-sewn opening. I was lucky to get a badge number that low; it had belonged to someone who had previously left to work for a neighboring department. I would find out later that the badge had been worn by Jim Tharp, whom I had played ball with and who would die very suddenly on a bike ride. Because he was a friend, that badge meant even more to me.

So, there I was, standing in the middle of my bedroom looking in the mirror at Buddy Martinette, paid firefighter, Badge #56 of the Virginia Beach, Virginia, Fire Department. Never mind that I didn't have the dark blue pants on yet, I just stood there in my underwear and that shirt. It was, to this day, one of my proudest moments.

Our recruit class was nothing like the recruit classes of today, although we were the first class in Virginia Beach to require an emergency medical technician (EMT) certification. Since a few of us already had our EMT certification prior to being hired, we were required to fill in for other firefighters who still needed their EMT training. My first assignment was not really a first assignment at all because

FIGURE 1–6. Virginia Beach Fire Department Recruit Class #13. Author on far right.

I was assigned to go from one station to the other to provide relief. Working 56 hours one week and 40 the next, it was a very confusing time.

It was so confusing that on my first regular assignment day, I called someone in the administrative department named Mrs. Cartwright to get clarification on my leave totals. She had instructed us to feel free to call if we ever had any questions about leave or pay. Turns out she really didn't mean that at all, and we had a less-than-productive conversation.

I Have a Question

District Chief Brammer showed up at the station. He asked me to come in the office and then very calmly inquired, "What the hell did you say to Mrs. Cartwright?" I told him the story and he proceeded to place me in the car and head downtown.

There I was, standing in Deputy Chief Quist's office as he proceeded to ream me up one side and down the other. During all the screaming, it occurred to me that he didn't have the story quite right and so while he was taking a breath, I tried to interrupt him to set the record straight. Guess what? The deputy chief of operations doesn't need to have the record set straight by a recruit firefighter. Another valuable lesson learned (fig. 1–7).

So I got fired in my fourth week of recruit class. I left his office with tears in my eyes, clear that my dream of being a paid firefighter had come to a very sudden end.

Chief Brammer, who had waited for Chief Quist to finish his tirade, asked me to wait outside the office for a few minutes and then he would give me ride back to the station. After an eternity, the door opened and Chief Brammer emerged. I remember very clearly; he placed his hand on my shoulder to offer some sort of fatherly gesture for my misfortune. As a chief, never underestimate the power your touch on the shoulder of another person has, as it speaks to love and caring.

Once we got in the car, I really lost it. Chief Brammer had known me since I was 15, and I very much looked up to him. I can only imagine how pitiful I looked sitting in the passenger seat of that car. After a few minutes, he just looked over and said, "Relax, you're not fired." Chief Quist had a fairly long list of folks whom he had fired and then unfired. I ended up on a distinguished who's-who list in the department of those who would fall upon that same fate. I came to admire Chief Quist. He was a fine man and a great operations chief—short fuse and all.

Finally, after EMT training, Recruit Class #13 got to the fire training. Back then, recruit class ran only slightly over 12 weeks long. On top of that, it took place in an old military fire station at Camp Pendleton in Virginia Beach. The old fire station was cold, and in the middle of January most of us had to sit in class

FIGURE 1–7. Deputy Chief Fred Quist (right) had a short fuse but was well respected by the author. Also pictured are Harry Diezel on the left and Frank Mixner in the center (courtesy of Ray Smith).

with our turnout gear on. At night, when the oil-fired boiler would finally catch up, the result would be soot-covered desks in the morning.

I finished recruit class just like I finished high school—middle of the pack, nothing special academically, although I felt I had good firefighting skills and could handle myself on the fireground. The trouble was that I now had the reputation of a troublemaker. Not only was I from Thalia, but I was also among the number of prestigious folks who had been fired, even if it was for only a few minutes.

Taking a Chance on a Troublemaker

My first assignment was Seatack #12 station, and our battalion chief was Jimmy Carter. Chief Carter was the quintessential chief officer. He always looked sharp in a perfectly pressed uniform, and he had great command presence both on and off the fireground. Chief Carter drafted me out of recruit class. Although he would be the first, Chief Carter would be one on a long list of people who would take a chance on me. Chief Carter retired as the deputy chief of the Virginia Beach Fire

Department and passed away suddenly in 2023. He remained a great mentor and friend (fig. 1–8).

The rest of my initial years in the fire department were filled with great fires and even greater relationships as a boot firefighter. I loved riding backward (in the jump seat) so that whatever happened I would be in the middle of it. While I never wished harm would come to anyone, if something was going to happen, I wanted it to be my engine company and me who got dispatched.

I spent the better part of my initial firefighting years enrolled in the 10-year associate degree program. The system back then was set up to reward firefighting and technical competence, not necessarily leadership capabilities. Study hard, do well on the fireground problem, and have decent evaluations, and you are an officer. Through this track, I became a lieutenant (fig. 1–9).

As a first-line supervisor, I was in my element. We ran fires, trained hard, and worked diligently to refine our craft. I also made my share of mistakes. I am forever grateful to the many men and women who tolerated me and my headstrong, get-it-done-at-all-costs leadership style. I would come to find that I had much to learn about myself and the leadership styles that would be most effective in encouraging and motivating people.

FIGURE 1–8. The author (left) becomes a new recruit at Virginia Beach (VA) Fire Department with Gary Weidner, Dan Fentress, and Greg Wakman.

FIGURE 1–9. The author, second from the left in the back row, gets a promotion.

The Battalion Chief Job That Got Away

After 12 years in some of our city's most active stations, I decided to throw my hat in the ring for battalion chief. I do not really think I had my heart in it. I enjoyed my time on the engine and ladder truck, and I was wondering what was the point of being a firefighter if you couldn't fight fires.

I did try for the battalion chief position anyway, but I did not get picked. It was a crushing blow because it was the first time in my career that I wasn't successful at achieving a goal, even with as little effort as I had put forth. I didn't want it that badly until I didn't get it. Then, something changed. I was, for the first time in my career, feeling the sting of rejection and failure.

I was so sure that I was better suited for the position than the other candidates that I went to see the chief. Chief Diezel is a great man and was a firefighter's chief. He was the kind of chief who not only knew you but also knew your wife and kids. Back then, the Virginia Beach Fire Department felt like a family fire department.

I didn't want to come off as arrogant because I didn't want to appear weak in the chief's eyes. His experience would tell him otherwise. His next words will forever be etched in my mind: "Buddy, you are a great firefighter and an outstanding company officer. The problem is that I need a chief, not an outstanding company officer." Talk about a reality check. I had some soul-searching to do.

I shared my disappointment with my father. He suggested I go back and find out what it was the other guy had that I didn't. At the very least, I would have some idea of the kind of chief the department was looking for in a first-time chief officer. So, that is exactly what I did.

It turned out that the other guy Chief Diezel picked over me was the better choice. R. B. Alley, who was stationed with me at the Oceanfront Station #11 at the time of the promotion, had accomplished much more professionally than I had. R. B. had a bachelor's degree, while I was still stuck in my 10-year associate degree program. He spent time in training and the Fire Marshal's Office. He attended the National Fire Academy and participated in many department committees and work groups. In the final analysis, it became perfectly clear to me that while I was working for the Virginia Beach Fire Department, Chief Alley was working on his career. He was making the sacrifices that, up to that point in time, I was not willing to make. He was a very well-rounded fire department employee.

I'm the guy who, by luck and God's good graces, had managed to find his way in life. The problem is that you can only go so far without putting in the hard work that is required to be successful at the higher levels of any profession. If you want to be successful, you must work very hard. Sacrifices must be made, and the sacrifices, it turns out, are many.

Someone Lit a Fire

Over the next few weeks, I made a list. Instead of just picking things randomly out of the air, I started by writing a fictitious résumé. I started with what I aspired to be at the end of my career and then worked backward, describing what would help me get there. When I was done, I had a résumé that would make even the most aggressive overachiever proud and inspired the overachiever in me.

I started with this: If you're going to do this "chief thing," then what kind of chief do you want to end up being? I want to be the chief. If you want to be the chief, what kind of education do you need? I need a master's degree. If you want to be chief, what kind of experience do you need? I need code enforcement and training experience. If you want to be chief, what kind of attributes are "chiefly"? I need values and traits that will help guide my personal and professional life. What makes chiefs successful? They have a strong network of friends and professional alliances.

One other powerful tool was to sit down and write my obituary. Your résumé speaks to your accomplishments, but your obituary speaks to what kind of person you ended up being. The actual way you conducted your life and the values and traits that define you as a human being. In the end, it's the measure of you as a person and, if you can figure it out before you die, when the inevitable day does come you will have added something powerful to the world and it will not have a damn thing to do with your accomplishments.

For the first time in my life, I aspired to be something. I had a vision for what I wanted to do with my life and the things that were required for me to be successful. I learned the words "desired outcome." I could state for the world my own personal values and dreams. This was the point of commitment!

When I had both my résumé and obituary completed, I worked on my list. I first put a transfer in to go to the Fire Marshal's Office. I didn't get picked for that job either. I put a transfer in for training, and bingo, my transfer was approved. That move turned out to be one of the best moves in my career, as I was able to get my training certifications. It also introduced me to the National Fire Academy. During my time in training, I studied course development; developed problem-solving, organizational, writing, presentation, and time management skills; and learned video production, technical rescue (State Heavy & Tactical Rescue Team and FEMA US&R Task Force Two), and, maybe the most important aspect of my efforts, the power of networking.

I finished all my degrees in 5 years, while at the same time working in various divisions within the fire department. It was my goal to get every state and national certification I could, and those processes started with instructor credentials and eventually led to the Executive Fire Officer Program. I ended up spending so much time at the National Fire Academy folks weren't sure if I worked there or was a student.

The sacrifice part comes from managing conflicting priorities. That's something that every great leader needs to figure out. Family, work, friends, social organizations, church, and hobbies are all just a few of the many priorities in life that don't stop while you're running headstrong toward your goals (fig. 1–10).

There came a time when I had to make some tough choices there as well. I loved to play softball and to golf. Both sports are time consuming and require constant practice if you want to be any good (in my case, just a little better than average). These two endeavors consumed the better part of my free time. There will be more on that subject in a moment.

Library-Land

By this point, Chief Diezel had retired and our Operations Chief Jimmy Carter had assumed the acting chief position. I could sense that Jimmy wanted to get the department moving forward, so he solicited the help of a library executive who had previously facilitated other work in the department. The plan was to get

FIGURE 1–10. The author's promotion to battalion chief

a group of department stakeholders together and put some words to paper about what it was we wanted to be when we grew up.

Chief Diezel was very good at benchmarking our department against others around the country that we considered innovative. Folks called us Phoenix East because Chief Diezel and Chief Brunacini, from the Phoenix Fire Department, were good friends. We established an exchange program where our folks went out there for a few weeks and the Phoenix Department reciprocated by sending some folks our way. It was a great program, and whenever we were able to spend time in Phoenix, we would come back with great ideas on how to either establish new initiatives or improve our operations.

Sometime during my training assignment, I was selected to participate on a strategic planning team. At the time, I was managing our video production unit and officer in-service program and helping with the company live burn evolutions. I had a full plate.

I was late for our first strategic planning meeting, and when I walked in the room people were sitting around in a half circle while our facilitator moved little characters around on a desk. After a few minutes, I realized that these little figures that had everyone's attention were Wizard of Oz characters. Wizard of Oz? What the hell kind of meeting was this?

I had the fight-or-flight response hit me and was challenging myself not to follow through on my first instinct, which was flight. How could a librarian understand anything as complex as a fire department! Doesn't this woman know I am a firefighter? I save lives for goodness' sake, and I don't have the time or inclination to play this game. I mean, really, all she does is books in/books out!

I was missing some gene that caused everyone else to be so engaged in this playtime exercise. I didn't even feel good in my own skin. As she continued talking, I found myself so annoyed by what was taking place that all I could hear was background noise, wah, wa, wah, wa.

I survived that first meeting, although I could sense that our facilitator could feel my increasing anxiety. She broke out a plan, and I like plans. Plans have starting and ending points with objectives that can be achieved. We've got something here to kill and eat. You'll hear a great deal more about "Kill It and Eat It" later. I've come to realize this is the fire services' favorite problem-solving technique.

Because of her insistence that I stay engaged and the enthusiasm with which she presented the information, I gradually bought into the process. The woman from library-land would have a great impact on my life. In the months and years ahead, I became more and more impressed with her vision and problem-solving skills. It fascinated me how she could look at a problem from such a different perspective. On most occasions, I would be seeing a problem and she would view the same situation not only as a different problem but also with a different solution.

Deborah Dunford was brilliant even if she was from library-land. Over the years, she has helped me with strategic planning, organizational development, facilitation, human resource issues, and much more. A great friendship between us developed as she mentored and coached me to see things differently and apply new techniques to solve organizational problems. She is a dear friend and one whom I must thank for many of my achievements as a leader. I would come to realize that Deborah, and library-land, were much more than books in/books out (fig. 1–11).

Special Operations

While working in training, I had the pleasure of working with Mike Brown. Mike was a phenomenal instructor and one of the best rescue technicians in the world. Mike introduced me to technical rescue and took me under his wing. He was also instrumental in helping me become a member of the State of Virginia Heavy & Tactical Rescue (HTR) Team. The HTR team was comprised of the very best of the best Virginia had to offer in specialized rescue program development and instruction. During my years on the road as an instructor, I had the pleasure to work with some incredible talent while teaching confined space, trench rescue, rope rescue, and collapse rescue operations.

FIGURE 1–11. Deborah Dunford, dressed as a character from *The Wizard of Oz*, was a mentor and coach.

The problem for me was conflicting priorities. Do I get off shift and go to the golf course or go help teach that rope class? Do I stay home this weekend and play in that ASA softball tournament, or do I travel with the HTR team and teach confined space? Questions like these came up weekly as I worked to do all the things I wanted and still be all things to all people. I very clearly remember the day that Mike Brown said, "You know, you really need to make a choice. Do you want to play games or be involved in technical rescue?" I suppose you can guess which way I went on that one and had Mike to thank for that (fig. 1–12).

The Virginia Beach Fire Department's technical rescue influence was not just local. Our instructors traveled all over the country teaching technical rescue operations. When the Federal Emergency Management Agency (FEMA) started developing the Urban Search & Rescue Program, our department was uniquely positioned to be one of the first teams. Our special operations chief at the time was Chase Sargent. Chase is the kind of guy who just gets things done. You don't always want to know how things got done, but nonetheless, they got done. Chief Diezel was brilliant in giving Chase just enough rope to be innovative but not enough to hang himself.

Chase was the driving force behind us becoming FEMA VA Task Force II. The 20 or so original teams were made up of 60 rescue specialists. The program

FIGURE 1–12. Mike Brown was a great instructor and introduced the author to technical rescue.

required teams to be self-sufficient for 72 hours and included doctors, paramedics, search specialists, heavy riggers, logisticians, hazardous materials experts, heavy equipment operators, and command personnel. I started in the program as a rescue specialist team leader and worked my way up to operations officer. Talk about opening doors.

Because of the people I knew, primarily Chase, but also others whom I had deployed with and worked with over the years, I was asked to help develop part of the National Rescue Specialist training curriculum. My assignment was to work with Bob Zickler on heavy moving and lifting. Bob was the deputy chief in Indianapolis and had already done extensive work in the lifting and moving discipline. Over the next few years, I was able to work with some of the best in rescue operations. Ray Downey, John O'Connell, and Don Shaver are just a few. These people represented the best of the best in their fields, and that presented me with extraordinary opportunities.

When you work with great fire service leaders, you position yourself to assume one of those roles. That's exactly what happened when FEMA started developing the Urban Search & Rescue Incident Support Teams (ISTs). The IST was alerted to respond as the command-and-control element when multiple task forces are deployed on an incident. The teams consisted of people whose jobs mirrored those on the rescue teams, although there were only two people for each job. I was selected to be on one of the ISTs as an operations officer. There is nothing better than being afforded the opportunity to operate on disasters of local, state, and national significance when you are a special operations guy. Later I will talk more about the lessons learned while operating on some of our nation's most significant disasters.

During the later years of my involvement in the state HTR team, Chase Sargent, Mike Brown, Dean Paderick, and myself started a company called Spec Rescue. Dean was the original state team captain and one of the first technical rescue operators in the history of special operations. Mike was the rope expert and finished a book on rope rescue. Chase was the confined space expert and wrote a book on that topic and was also the company expert in trench rescue and my mentor in that field. I ended up writing a book about trench rescue operations, although I surely owe Chase the credit. He taught a guy like me with limited craftsman skills how to teach a discipline that is very craftsman oriented (fig. 1–13).

Code Enforcement

Code enforcement was not something that I was excited about. I mean, really, how can the Fire Marshal's Office compete with the Murrah Building bombing?

FIGURE 1–13. The author credits Chase Sargent (center) with teaching him the ins and outs of trench rescue.

Nonetheless, code enforcement experience was on my bucket list, and I knew I needed to get that part of fire service experience. Remember when I didn't get picked for the fire marshal job the first time? Turns out that it was a blessing of sorts as I review the history of my work experiences. The next 3 years as a fire marshal were in fact the most important in my career development plan. I learned more about the fire department mission, politics, team building, and organizational skills than just about any other time in my career.

My First Chief's Job

I was finishing my last Executive Fire Officer class at the National Fire Academy when I saw a posting for a chief's job in Lynchburg, Virginia. Like most of you, I skipped over the fine print and went directly to the salary line. *Crap, I make more than that as a battalion chief,* I thought, so I disregarded the flyer and returned to class.

A few months later I saw the same flyer, although the salary figure had been adjusted to what someone would reasonably assume is a professional fire chief's salary. I grabbed the flyer and took it home for a more intense review. Many of the requirements were in line with the objectives I had already achieved; executive fire officer, master's degree, varied fire department experience, budgeting, strategic planning, and change management were just a few. The bottom line was that I realized for the first time that I could compete for a chief's job even if I still had reservations about my abilities to do well in the position. I also realized that all those little incremental achievements had in fact accomplished their objectives. I was as close to being chief material as I had ever been.

Filling out the application was somewhat of a personal celebration. I had accomplished some things over the past 10 years, and seeing it on paper gave me great satisfaction. With the résumé and application complete, it was do or die time, and so in the mail it went. Then there was nothing.

After a few months' time I was mailed a series of questions, most of which dealt with leadership and personnel issues. I completed and sent in my responses and was scheduled for a phone interview with the Gallup Organization. I was then sent over to a local college to do a video interview. Then there was nothing again, and when I say nothing, I mean nothing. It was as if my life's work on paper and all the other efforts just got sucked into a black hole.

Not wanting to appear anxious, I didn't call or write. I just let it fade from my mind as I wrongfully assumed I just didn't measure up. As it turns out, it's not unusual to never hear back about your status in a chief's hiring process. I personally think this is very unprofessional, and as the chief of my own department,

I choose to notify a candidate, good or bad. In our system, if you're moving on, you're told, and if you're not moving on, you are also told by being thanked for your participation.

A few months went by, and my wife, Sarah, and I were sitting on the front porch of our Chesapeake home when the mailman made his daily delivery. I opened what I thought was a "Dear John" letter from the Lynchburg Department of Human Resources. Instead, it was an invitation for three of us to participate in a chiefs' assessment center the next month. I wandered off toward my truck, which was parked in the driveway. "You're going somewhere?" Sarah asked. "No, just getting a map to see where Lynchburg is," I replied.

The assessment was made up of a PowerPoint presentation where I was tasked with creating a public safety educational facility. The interview with city department heads was mostly about strategic planning and my incident management experience. The interview with the city manager and his staff mostly pertained to code enforcement.

It was almost like everything was scripted based on how I had prepared over the last 10 years. The résumé was complete. The education was complete. The video interview was very comfortable for me because of my time in training. The presentation was spot on because of my instructor experience. I had nailed the leadership and incident command information because of my varied local, state, and national rescue experiences. The first 6 out of 10 questions in the interview with the city manager were about code enforcement.

When I left the last interview, I knew I would be hired. The city manager and I connected, and he reminded me of my grandfather, which put me completely at ease. It was like Granddad said, "Here's one for all your hard work, son" and then placed a man in front of me that acted and looked just like him. That night, in the hotel lobby of the Holiday Inn, City Manager Charles Church offered me my first chief's job.

The week before I was to start my new job, I received a call from Chuck. It was not good news. He had been diagnosed with cancer and only had a year or so to live. On the phone that night, I could hear in his voice the commitment he would make in his last months to ensure I would be successful. Chuck worked another year before actually leaving the job. He lived a few months longer. It was a very sad day indeed when my friend Chuck Church died. In the end, he was true to his word and gave me all the support I could have possibly wanted (fig. 1–14).

During my first year in Lynchburg, Chuck encouraged me to attend a program at the Senior Executive Institute at the University of Virginia. Now, it is mighty hard to get a diehard Virginia Tech Hokie fan excited or even forcefully ordered to participate in anything that is associated with or located at the University of Virginia. All kidding aside, I was lucky to get accepted to this program, as it was primarily designed for city and county managers and their assistants.

FIGURE 1–14. Chuck Church (left) encouraged the author to pursue city/county leadership opportunities.

With the human resources director in Hanover County, I attended the two-week course and then in subsequent years the leadership updates the program provided. Listening to city managers talk about the challenges of local government management provides an interesting perspective. I think that has helped me be a better fire chief, and at the very least, I made an incredible number of local government management friends. Did I mention the power of networking?

City and County Management

I maintained my relationships with many city and county government managers over the next few years, and frequently they would call me about fire service issues. On more than one occasion, they suggested that I try my hand at a leadership post that was a little more global than just being a fire chief. The thought of effecting leadership at higher and higher levels appealed to me at the time, and so after 6 wonderful years in the Lynchburg Virginia Fire Department, I staked my future on another application dropped in the mail.

Eager to get closer to my birth home in Virginia Beach, I applied for a job as an assistant county administrator in Hanover County, Virginia. You will hear a great deal more about that adventure later in the book, where I devote chapter 24 to chief and manager relationships. The bottom line is that I was afforded an opportunity to learn from another great local government manager. Cecil "Rhu" Harris turned out to be a real local government management professional. As the county administrator, he was as sharp a guy as I had ever met. It seemed to me that he never forgot even the smallest of details, and he was brilliant at reading and managing elected officials (fig. 1–15).

My job was to manage Fire and EMS, Emergency Communications, Building Inspections and Code Enforcement, and Animal Control. Just as a side note, the folks in Animal Control are some of the best people in the world. Their customers, because of the emotional attachment people have with their pets, are almost always a handful. I learned a great deal from the various department heads who worked for me. All of them were extremely committed to providing a great product along with superb customer service.

I didn't like being number two. It was not that Rhu was a bad leader—that couldn't have been farther from the truth. Some people are cut out to be good number two guys, and some are not. I found that my leadership style was very much about seizing the leadership moments when they occur and then acting on them to form great relationships. When you work for someone else and that part of the job is their responsibility, you need to be careful how you apply your own leadership. I didn't much care for the "careful" part.

At the end of the day, and with rapidly approaching retirement eligibility, I could feel the tug of the fire service. I missed the emotional aspect of what we did in the fire service and the bond that is created between members of the fire

FIGURE 1–15. County Administrator Rhu Harris was a political tactician.

service family. With those feelings in my heart, I would soon find myself back in the fire service.

Back to the Fire Service

After a stint as an assistant county administrator, I relocated to be the chief of the Wilmington, North Carolina, Fire Department (WFD) (fig. 1–16).

Similar in many ways to Lynchburg, it is a department steeped in history that began in 1846. The WFD is a great fire department, and those men and woman worked hard to become a "world class" fire department. They challenged themselves to be the fire department that other fire departments look to when things like innovation and forward thinking are discussed.

I was hired by a city manager named Sterling Cheatham, another great leader and local government manager. Sterling was the finance director for the City of Norfolk when I was a battalion chief in Virginia Beach. The two cities are next to each other, and we both enjoy discussing much of what has happened there over the years. He is also a graduate of the Senior Executive Institute.

After a long and satisfying career in public service, I returned home to be the chief executive officer of Spec Rescue, a rescue and specialized operations

FIGURE 1–16. The author as fire chief of the Wilmington, North Carolina, Fire Department with Deputy Chief Steve Mason (left)

consulting company that I helped start some 30 years ago. Leadership in the private sector is just as important and relevant as it is in the public sector, and in that regard the principles in this book will apply to both.

So that is me in nutshell. I have been afforded the opportunity to work in a great profession with great people. I am blessed for the many people God has placed in my life.

Life Lessons Learned

Lessons I learned from my childhood and those taught to me by many mentors all add up as important occurrences in my life.

As I close in on 65 years of life, 45 of which have been spent in or around the fire service, I believe that I am only just starting to fully understand leadership and the power of effective leadership.

During strategic workgroups I have facilitated over the years, we have defined *culture* as the collective experiences, traditions, and values that shape behavior in an organization. I believe this same definition can be applied to leadership, as somehow great leaders not only experience traditions and values, but they are also able to use them to change culture and lead people toward excellence. I can only hope that this book helps you realize your leadership potential and then lead people toward excellence.

2

The Old Man and His Flowers

Leadership is like a bowl of soup being served to people who all have different tastes. It takes many different types of ingredients, all somehow used at the same time, to be a great leader to all people.

Key Points

- The totality of you is created out of the many experiences you have.
- Employees come to us desperately needing to be fertilized to grow.
- Bad employees can ruin good ones if we don't take time to weed them out.
- Leadership is first and foremost about people.
- Effective leaders must create a nurturing environment to be successful.

When I awake, it is to the sound of water running through baseboard radiators. The water is heated in a boiler located somewhere in the bowels of the house where no 7-year-old would dare venture.

By the smell in the air, it's clear Grandma is quite a way along with cooking somebody bacon and eggs.

I could usually find my grandfather sitting in a lawn chair under a very large oak tree just steps from the breezeway exit. It was a stately tree with large, low-hanging limbs just far enough apart to afford young explorers an opportunity to see the world from heights under those which no mortal dare goes for fear of a fall.

While I can't yet see my grandfather, I can tell that I am close because I can smell his pipe tobacco.

Jesse Godwin Futrell, a retired Norfolk streetcar operator and civil service bus manager, hailed from Conway, North Carolina. His parents were farmers, and they needed to be good at it to feed and support the 13 siblings who called the farm home. As was the prevailing wisdom of the day, when crops needed to be cared for, kids stayed home from school and farmed. School was a bonus that would be fit in around more important duties like planting and harvesting peanuts, cotton, and tobacco (fig. 2–1).

The farming and the on-and-off-again school participation ultimately meant my grandfather's formal education ended by almost graduating from the sixth grade, a fact I mention here because he was very proud of the self-sufficient way he cared for his own family. The bottom line though was that while he could read and write, it was only enough to get by, and by today's standards he would have surely been labeled as illiterate.

At 16 years of age my grandfather was kicked out of his house by his father. He would do odd jobs for the next several years until hooking up as a streetcar operator. When he formally retired from civil service, he took to the only other thing he knew how to make money doing, and that was to grow things.

FIGURE 2–1. The author and his grandfather, Jesse Godwin Futrell

So, there he sat. Before him was a 2 ft. high, 10-by-10 square box of concrete blocks that contained pallets and pallets of little seedling cups. Rising above the pallets was an irrigation system that, when started, produced a very fine mist of spray designed to keep the entire area moist.

I watched as he gently used his pocketknife to carve back the bark on a small twig that on one end was cut and from a beautiful red flower. After stripping the bark back some three quarters of an inch, he would set that flower aside and repeat the process with an equally beautiful white flower. He would then place the two ends together and use a small rubber band to hold the two pieces together.

At this point, he would place the rubber-banded seedlings in a combination of soil, peat moss, and cotton seed meal, the latter of which is used as fertilizer. The combination, in proper proportion I assumed, was instrumental in producing variegated red and white azalea blooms on a single azalea bush. When all the trays were filled, he gently placed the pallets under the series of pipes and sprinklers and then covered the concrete block housing with clear plastic.

"What are you doing, Grandpa?" "Makin' a bush," he said. "What kind of bush?" "A red and white bush," he said. "Why are you cutting the ends off?" "Because they will not grow together unless you peel the bark back," he stated. "How come you use rubber bands?" "So the flowers don't come apart." "What is that black stuff?" "It's called peat moss." "What does peat moss do?" "It helps the soil stay moist so the seedling can grow." "What is the yellow stuff?" At this point, he's becoming somewhat agitated with my questions, and I'm sure wondering why Grandma has not called me for breakfast.

With the patience of a saint, he goes on to say that the yellow stuff is called cotton seed meal and it's what feeds the plants and helps them grow together. "What are the pipes for, Grandpa?" "Isn't that your grandma calling you, son?" "Huh, I didn't hear anything!" "Well, the pipes are so I can water the seedlings without removing the plastic." "What is the plastic for?" "I am almost sure I heard your grandma calling. Aren't you hungry?" While I stare very intently at his right ear, he says, "The plastic helps hold in the moisture." "Why is it clear?" "It's clear so the sun can shine in and warm the flowers. Come on, let's you and me go get some breakfast."

And there you have it! To be a great leader, all you need to know is how to breed variegated azaleas as taught to a 7-year-old boy by a man who could barely read and write and who didn't graduate from the sixth grade.

Some years ago, when finally having the responsibility to lead my own fire department, I started pondering this whole leadership thing. Sure, I had read all the books and, up to that point in time, had turned most of life's lessons, good and bad, into learning opportunities. But really, what is the difference between people who are successful leaders and can manage the turmoil of changing conditions and those who just manage to maintain the status quo and ultimately get moved aside for someone who is seen as the new panacea of leadership?

It was about that time that I realized my granddad knew a lot about raising flowers, but he indeed may also have known a great deal about leadership. You see, each of us enters the workplace as a seedling that has been stripped from an adult bush and may or may not be healthy and vibrant. Vibrant or not, we enter the workplace and are grown along with the other employees in the organization.

The bottom line is that all leaders want their employees to grow from seedlings into beautiful and disease-free plants that can be placed with and add value to the other aspects of our organizational landscape. To accomplish this, we want to hire disease-free seedlings, give them some fertilizer, and then plant them in a healthy environment that will ultimately encourage future growth.

Unfortunately, we sometimes get healthy seedlings and place them in environments that stunt their growth or, worse yet, cause them to die. In still other cases, we hire diseased seedlings and wonder why they don't flourish in a perfectly acceptable and caring environment.

When a new employee comes to us as a seed or flower, they need to be potted to develop roots. Understanding that, we first gather this seedling (hire them) based on the values the organization holds as their own, so like the flower, we know the stock from which the cut came. We then take this very best employee and put them in formal training for their position so that they will have the fundamentals necessary to provide high-quality service to our customers.

To this mix, we encourage additional formal education and individual wellness counseling along with mentoring as given by our current employees (the fertilizer). We water the employee every couple of weeks by giving them a paycheck and then wash our hands and hope the employee continues to grow.

Even still, peril is around every corner for our new employee. Yes, even after these cuttings have produced roots during their initial training, there is no guarantee of success. As we remove them from their cups, care needs to be taken in deciding where they will be buried in the earth. The root ball must only be placed in the soil at the right height, and only the right amount of sun will be best for optimum growth. Then, occasionally, a little fertilizer needs to be added to generate new growth and health in the plant.

All of what we have said before this remains vital in the employee growing process. If any one aspect is in the wrong proportion, you could get an overeducated, middle manager with no people skills or a first-line supervisor who is a great technician but does not understand the larger, more global systems in which we operate.

Of course, we are really talking about people and not azaleas here. Each of these people comes from a different background and is raised with and taught a sometimes distinctively different set of values by their parents. The values are reinforced through modeling and experience.

For leadership, this is not a comforting position. How do we take all kinds of people with varied backgrounds, different values, and unique traits, and then align them toward a common cause, all the while expecting some semblance of consistency in personal behavior? This is no easy task, as many of us find out each morning when we open the office door, sit down at the desk, and then start the process of dealing with human beings and customer service issues.

What do leaders need to do to be successful? First, we can start by understanding that leadership in organizations is first and foremost about individuals. It is personal and based on relationships. That includes all individuals who make up the organization and therefore means us as well. All components of personal leadership need to be addressed if individuals are going to be compelled to take it upon themselves to act in a manner that produces great customer service. This is especially true if the great customer service requires sacrifice on the part of the employee. Growing and nurturing individuals is paramount to any successful organization.

> For leadership this is not a comforting position. How do we take all kinds of people that have varied backgrounds, different values, and unique traits, and then align them toward a common cause, all the while expecting some semblance of consistency in personal behavior?

Secondly, effective leaders must create a nurturing environment. Frequently, we call the effect of the environment the organization's culture. Culture is described here as collective wisdom and knowledge of the organization as represented by the behaviors that result. The environment is what determines if the seed grows and how well it flourishes to grow additional seeds.

To illustrate this point, I can tell you that employees' attitudes almost always reflect the attitude of the leaders and subsequently the prevailing culture of the organization. It really does not matter whether you are talking about safety, health and wellness, or concern for customer service. If the leader does not care and put forth effort, the employees will do the same. The bottom line is, you can take a good seedling and place it in a bad environment, and it will die or become diseased. The late Phoenix (AZ) Fire Chief Alan Brunacini said it best: "If you follow an ugly kid home from school, you can bet you are going to find ugly parents." I think that says it all!

Let's first talk about how to take care of ourselves and our employees and the direct relationship this has on our ability to create high performing organizations. If we take care of the inside, the outside will take care of itself.

Our journey will begin with a look at personal leadership. Personal leadership is the tool we use to build credibility so others will think it is worthwhile to follow us. Credibility is rooted in personal values and the traits that manifest themselves because of using these values to make decisions. Once people feel you are a credible leader, and they understand the values you use to guide your decision-making, you will need to provide followers with a vision.

Next, understanding that personal aspects of leadership are used to create healthy and vibrant employees, we need to create the culture that will most effectively secure long-term health, in effect creating an environment conducive to employee motivation and great customer service. Having a vision for what you want to achieve also includes creating an environment where it is safe for people to help you achieve the vision, with respect to allowing independent decisions free from ridicule and micromanagement and letting your employees act and behave like adults.

We will talk about putting the right people in the right place, with the right skills, at the right time—thus taking advantage of the individual strengths of our members to help the organization achieve success. We'll then discuss creating accountability and responsibility as an outcome of treating people like adults. Finally, we'll finish our journey with a discussion on relevance and how to stay relevant in our fast-paced world.

3

Understanding the Leadership Commitment

The higher I go in leadership, the more I realize that my job is about the art of knowing less and less about more and more.

Key Points

- Leadership only comes to us by way of followership.
- Great leadership is about heart in your people and what they are trying to achieve.
- Leaders must learn the art of managing conflicting priorities.
- The entire foundation of leadership is built on trust.
- There is no one leadership key that unlocks all doors.
- High performance leaders can adjust their styles based on situation and need.

Your leadership journey begins with an understanding of what it will take to be the leader who others aspire to become. This kind of commitment lays the platform from which all that comes from your efforts is realized. You will be individually judged on your leadership by your ultimate ability to create followership. Leadership resides in the mind of the follower. For leadership to take place, the follower must be willing to be led. The reason followers are so important is because, in the end, they influence the outcome of any initiative.

The very first thing I want you to know about leading high performance organizations is that the process is foremost about heart. It is about the motivation to lead and help others be high performing, driven by pureness of self and not by self-advancement. It is all about a commitment to elevate others and never about beating your peers to the next rung on the ladder. Being so dedicated to others that you always place them first and placing the importance of the mission above all else even if personal sacrifice is required constitutes servant leadership.

I have always been fascinated with the concept of servant leadership and how it is used to manage conflicting priorities. Much of what a high performance leader does is manage conflicting priorities. In any given situation there are winners and losers and the haves and have nots. It's difficult to manage conflicting priorities when the instinct of humans is to first preserve

> How can it be that all of us are trying to get to the same place and yet we mistrust how others play their own special roles in getting us there?

and protect themselves before moving on to how a decision might ultimately affect other people or the organization.

As a fire chief, I am frequently reminded by people not to forget where I came from, not that I am from Virginia Beach but instead that I am only a few years (well maybe a few more than just a few years) removed from riding in the jump seat answering fire and EMS calls. As I reflect on the wisdom of being grounded by my roots, it occurs to me that this happens each time we progress up the chain of command. Usually this is followed by the rhetoric, "Boy did he change" or "Man those bugles really went to his head." Translated, this means you are not one of the "us" anymore and that somehow your allegiances have changed, and you are now one of "them."

How can it be that all of us are trying to get to the same place and yet we mistrust how others play their own special roles in getting us there?

Many great leaders agree that leadership is mostly about successfully managing individual priorities that may or may not help achieve organizational success. For instance, we (the fire service) frequently advocate for more equipment and personnel so that we can do a better job of addressing fire and EMS issues. Consequently, we view all those opposed to this lofty endeavor as some form of adversary who must be struck down.

It then becomes an officer's task to mediate these priorities. But what is the basis of the conflict? Is it that we just want more personnel and equipment, and our bosses want us to have less? I would argue that if this is how you perceive the situation, you will always be in conflict over issues on which you have a differing opinion from someone else.

If all participants in a situation are not focused on the same issue, their perception will always be one of conflict. In the situation above, all parties should be focused on customer service and not the specifics of how the service is delivered. The problem is that the specifics will affect them personally before they move on to the merits of how effective the direction is at achieving its goal.

To be effective at managing conflicting priorities, we need individuals to buy in. No individual's self-interest should stand between the organization and the organization achieving success at its mission.

When we are questioning each other about a conflict in priorities, we first need to look at where the focus of the priority is. In situations where you are managing conflicting priorities between yourself and someone else, be sure both of you are focused on the same goal. Open, honest discussion with each other, along with a healthy dose of servant leadership mentality, makes us all richer in spirit and body. Nothing good comes from a situation when a public servant is focused on serving themselves and not the customer.

It's not that we change as we get higher in our profession or that our priorities change; it's just that our job functions change. We would do well to understand that and help each other stay focused on the true goal.

Retired FBI Agent James Reese said, "As leaders, we should wake up each day and ask what we can do for others; to always seek to take the spotlight off ourselves and place it on the people we work with and for." I would challenge all of you to look inside yourselves and question whether your daily actions and efforts are focused on your family and the people you work with or instead on shining the spotlight on yourself for your own good.

> To be effective at managing conflicting priorities we need individuals to buy in, and when I say buy in, I mean buy in to the point that self-sacrifice is just another aspect of achieving the goal.

Leadership begins internally, as no leadership takes place before a person learns self-leadership. Because self-leadership is first about heart, and most importantly about having the heart to know that it's about placing others first, it makes perfect sense that your effectiveness will be driven by the relationships you build with people and your commitment to help them grow and become better. This takes place through relationships that motivate and inspire people to forecast and respond to the future, actively manage current functions, and then ultimately provide meaningful direction for the organization.

Understand this: Some people will not like you. Some people will despise you because, as a leader, you are expected to make change and that change can be viewed as either positive or negative depending on how it affects them as individuals. That is not to say that people cannot elevate themselves to see the higher needs of the organization. Successful leaders are not the ones who are most liked or agreed with—they are the ones who soften the blow and inevitability of change and then actually achieve progress that benefits the employees and the citizens we serve.

Understanding relationships in the leadership business is critical because to be a successful leader you will on many occasions need people to act without specific direction. That requires trust, and trust is developed when people know they can depend on you. Most of the time you will be trying to take people to a place that they do not know how to get to and, equally as difficult, cannot see a reason for going there in the first place.

> Your success as a leader, and ultimately the success of your organization, is based on your ability to gain followership and then further the follower's ability to perform.

Trust also helps when there is an inevitable gap in our ability to communicate real-time progress with employees and our customers. Since communicating everything to everyone in real time is impossible, effective communication relies on the both the sender and receiver understanding the context of the communication in the same way. Even in the most successful high performance organizations, trust is necessary and will ultimately buy you time until you can close the communication gap.

Trust is important because leaders who want to create a high performance organization must create a learning organization. A learning organization shares information on successes and failures because doing so improves the likelihood that the entire organization will become better. A learning organization can only be successfully implemented when constant experimentation is encouraged. If you want this type of environment, you must trust people and be willing to accept that occasionally they will fail. When they do fail, trust will be the platform on which the next successful effort will start. Remember, how we learn as an organization and how we grow in effectiveness are inextricably connected.

The very nature of rules also says to someone that there is a boundary from which they are required to operate. Good leaders stop evaluating small details about everything and get rid of controlling mechanisms so employees can act on their own to solve problems. The next time you are contemplating writing another rule, remember that someone with no choices has no room to decide and therefore no room to display leadership.

Leaders have many different qualities and personal characteristics, including a dedication to servitude that devotes much of their life to a cause, all of which help make them effective. When we focus on servant, selfless, and dedicated leadership, in the end, we regard the effectiveness of a leader by the relationships they created and the lives that they touched.

Leadership success in high performing organizations can loosely be tied to a combination of unquestionable values, intelligence (the capacity to think logically about a situation), and emotional intelligence (the ability to manage one's own emotions and the emotions of others). The undoing of many leaders is that intelligence gets them in the room, but lack of a decent personal value system and inadequate ability when it comes to emotional intelligence limit their abilities once they are there.

If you buy into the fact that relationships, values, intelligence, and emotional intelligence are important to your success as a leader, then perhaps the next thing you should realize is that leading is also hard work—24/7 kind of hard work. As a leader, you are expected to continually evaluate your organization's effectiveness, work to remove key obstacles preventing further effectiveness, gain the support of key people to assure effectiveness, and then make sure your organization has the necessary resources to remain effective. If you don't do this, there will not be any "world class" or "high performance"; rather you will just live to see another day.

There is not one key that unlocks all doors. That is what makes leadership such an interesting practice. To be effective, a leader needs to recognize that leadership is many things. It is in many ways like combining ingredients to cook a meal. Some ingredients work better when mixed with others to make certain dishes. You can take a bunch of good ingredients and mix them together in the wrong proportion, and you will end up making something that nobody wants to eat. The ingredients could all be fresh and tasty; however, the pro-

> How we learn as an organization and how we grow in effectiveness are inexplicably connected.

portion for one of the ingredients may be off or some of the ingredients in combination with each other may create a bad flavor. If that's not daunting enough, in addition to knowing how much of each ingredient is important, you also need to understand when to add the ingredients to the pot. Timing in cooking, just like leadership, means that some ingredients need longer to simmer than others. In this example, not every leadership ingredient cooks at the same rate. You also need a kitchen full of tools and appliances to make your meal turn out to be worth eating.

Leadership opportunity in crisis presents itself when a vacuum is present and people are looking for someone to step up and tell them how they are supposed to behave and feel. The most effective leaders in a crisis live in the moment with their followers. They engage in the situation by being down "at the street level." Remember, the leadership opportunity in a crisis occurs at a time when people need some sort of symbolism. They are looking for the leader to tell them how to resolve the situation, how they should feel about the situation, and then how they need to behave while it is being resolved.

There are two sets of very important steps you will take as a leader if you want to create and establish a high performance work environment. The first ones you take on the journey will define who you are and what you represent. Then, the last few steps you take put an exclamation point on your career and say to everyone that you had the wherewithal, discipline, and courage to see things through to the very end.

Several years ago, I watched the San Antonio Spurs dismantle the Miami Heat in the 2014 NBA Finals. What was remarkable about this five-game series was how much each individual player on the Spurs had complete and total buy-in on the system. While each player was individually accountable for his contribution, it was their contribution within the system that produced remarkable results.

The lesson in this example is that your employees must understand your system, collectively believe they have a contribution to make in the system, and have an appreciation for what the system can achieve. Any commitment to a system will not succeed without the full support of all your workers. You will need to fight like hell to stay the course and see the process through to the end. It will seem easy on some days to just give in and submit to the small but vocal minority whose only motivation is to break down the system.

> Confidence, in the leadership business, comes with experience and knowledge, and both of those attributes can be at a premium at the beginning of your career.

What causes people to not buy into your system and to actively fight against you and the organization? First, when a system is developed, specific accountability is assigned for how members contribute to the success of the system. Poor performance on the part of one or more employees means the ball does not get passed around and the system breaks down. Individual success becomes more important than organizational success, and the observation of where the breakdown occurs will be plainly evident. The bottom line is that some folks are very comfortable operating in a system where they can pass accountability for success or failure onto the shoulders of another team member.

It will be important in your very first leadership years for you to understand that the playing field will never be level when it comes to you and your ability to defend your leadership qualities. Those people who don't care for you will be able to say anything because when they speak about your leadership, they can do one of two things: say something true about you or make something up. In either case, defending yourself can be difficult, and the way you choose to defend yourself is in and of itself a defining leadership quality.

For some folks the first few years are the hardest. These will be the years when confidence is at its lowest point and can be easily shaken by team members who, just like kids on a playground, can be cruel and ruthless. In the leadership business, confidence comes with experience and knowledge, and both of those attributes can be at a premium at the beginning of your career.

When you endeavor to become a leader, it's important to understand that many of the people who want to crack your leadership confidence are doing it because it's human nature. Good and bad, yin and yang, hot and cold, it's the differences in us that bring about organizational balance. Always remember that the folks who don't care for you are providing a most important service. First, their services are provided free of charge. Second, each time they say something bad about you and it turns out not to be the case, you have just benefited from a credibility boost that you could have never had enough money to buy.

One of my Executive Fire Officer instructors told me very early on in my career, "People want to be led, and then will resent being led." What will define your leadership skill over time is the commitment to display the same pattern of behavior over the long haul of your career. Do the right thing, live by your organization's values, and then let people decide for themselves.

At some point in your journey, you will begin to build confidence in yourself. Just like a baseball player on a hitting streak, the ball will seem bigger as it approaches the batter's box and those routine fly balls, for some reason, will one day not be caught but rather fall in an open space on the field. For some, this is the highlight of their career. It is also at this point when those confident leaders feel all in the world is great and they have "made it."

For all my officer in-service classes, I save time at the end to meet with the various levels of the department. Recently, I met with my captains to answer questions and address issues in an open forum where just about anything could be shared. This kind of situation can be very uncomfortable for some leaders, as there is little protection from some comments that are generally based on perception and not fact. That said, you know what they say: where there is smoke, there is always fire.

During one of these meetings, one of my captains said, "Chief, when are we going to take a break?" Trying to first understand his point I thought, *What break?* When does life for successful people ever become about a break? Don't successful people continue to push themselves to be better, learn more, and be more productive? Can't the same be said for successful organizations?

So, taking a break is bullshit. Life does not offer a break. There is no stopping, rather only moving forward. If you stop wanting to be better and to continually improve, it will signal the end. If you're tired, then step aside because a successful leader who decides they will rest on their laurels and ride out the rest of their careers is no help to their organization.

One of the things I challenge myself to do at the gym is to have a little something left in the tank at the end of my workout. Whether it's running at 6 miles an hour for 27 minutes and then 7 miles an hour the last 3 minutes, I always want to have my best at the last. The same can be said for leadership.

A manager hires a fire chief because they have certain traits and character-istics that are perceived as necessary to move the organization forward. As the leader works in the organization and applies those traits, the organization starts changing and therefore the needs of the organization change. The most successful and tenured leaders are those who can adjust their leadership styles to meet the needs of the organization at any given point in time. This lends itself to the concept that the leader's years of service in an organization may be related to their ability to adjust leadership based on the needs of the organization and not their own personal preference or style.

> The most successful and tenured leaders are those that can adjust their leadership styles to meet the needs of the organization at any given point in time.

The point to this is that the last years are just as important as the first. Your dedication, discipline, and enthusiasm to change will be monitored each day by the people you are leading. If you feel yourself getting tired, don't slow down but rather speed things up. If it turns out you're exhausted and don't have the energy, get out and find something else to do that will once again create that inner flame all of us need to be successful. Your people deserve a committed and enthusiastic leader from day one all the way until you or someone else decides it's the last day. Just remember this: as a leader, you will mostly be remembered for the discipline you had to hang in there and run hard right to the very end.

Principals of
High Performance

The process to continually improve is sustained only when there is a personal commitment to self-awareness and constant reevaluation.

Key Points

- Leadership in organizations takes place at all levels.
- Written outcome statements are essential to high performance.
- Outcomes should be written with an action verb and an "in order to" statement.
- Outcomes are useless unless you can measure progress in achieving them.
- This chapter presents 11 concepts and principles of leadership.

What comes next are the 11 most effective principles regarding high performing organizations. I will provide you with a tool to assess your department's, division's, section's, or work unit's current characteristics when it comes to high performance.

You need to recognize that leadership in organizations takes place at all levels, departments, and work processes. The expectation is that all of us will be leaders. Employees in positions of formal leadership should in

> Employees in positions of formal leadership should in all situations foster a work environment that encourages individual greatness and establishes a benchmark for continuous improvement.

all situations foster a work environment that encourages individual greatness and establishes a benchmark for continuous improvement. To that end, all of us are leaders, and the principles discussed here can be applied to every facet of our organizations.

Outcome statements written for high performing organizations are written with an action verb and an "in order to" statement. This is critical to successful implementation of any strategic task or other performance outcome.

Each outcome statement starts with an action verb. Action verbs describe an action on the part of some person or thing. Each of the statements listed in this chapter starts with an action verb. The second thing about a proper outcome statement is the action verb directly precedes the articulation of the tool you will use to accomplish your objective. For instance, in the first objective, it says to utilize (action verb) organizational values (tool).

After the action verb and tool are written, they are followed by the statement "in order to." The "in order to" do something is the most critical element of the objective. For instance, a solid outcome statement is to utilize (action verb) organizational values (tool) "in order to" promote consistent decision-making (desired outcome).

The point to writing outcome statements this way is that it keeps people on track. When things get hectic in committees and people's personal agendas creep into the work product, someone like you can say, "Wait a minute, this objective says that we are doing this 'in order to' accomplish this." If you have agreed upon this desired outcome at the beginning of your efforts, the logic will be hard to argue.

Outcome statements, in general, are not worth their salt unless you can measure how effective you have or have not been at accomplishing them. The magic here resides in being able to measure the desired outcome and not the tool when assessing progress toward achieving your goal. For instance, using our example of utilizing organizational values to promote consistent decision-making, many leaders would establish a survey that asks employees how well the organization operates with respect to utilizing organizational values in decision-making. The information received in response to this inquiry would only tell you how effective you are with using the tool, which is organizational values. The results in this example would not tell you how close you are to achieving your desired outcome, which in this case is to promote consistent decision-making. The bottom line to all of this is that your assessment tools need to measure how effective you are at achieving your desired result, which is what follows the "in order to" statement.

These specific concepts and leadership principles, when implemented, will help you create a high performance organization. These main objectives, and the many stories and examples of how they are implemented, are woven into the words and chapters that follow.

Please note that the concepts listed here are not the only factors that contribute to high performing organizations, but they can be considered a great start. I have included a survey in appendix A that will help you measure your effectiveness as it is described by your followers. The assessment will, at the very least, provide you some insight on where your organization is strong and where it may be weak. In this way you can benchmark your current reality and provide a starting point to measure future progress in achieving high marks for each of these outcomes.

1. Utilize consistent, published, and defined values in order to promote consistent organizational decision-making.
   - Understands rules cannot be written for everything
   - Provides a framework for employee decision-making in the absence of established policy
   - Sets an expectation that decisions outside of policy need to be aligned with established values
   - Discourages the employees' use of personal values when dealing with organizational issues

2. Develop and nurture quality personal traits that can be patterned over time in order to demonstrate consistency in leadership behavior.
   - Provides consistent leadership
   - Provides a platform from which consistency of decision-making can be expected
   - Makes certain the speech and actions of your behavior matches

3. Coach, mentor, and lead by example in order to build relationships and promote mutual respect.
   - Builds trust and shows employees that you care
   - Shows that you consider everyone to be in it together
   - Provides a framework for quality 360° feedback
   - Contributes to mutual accountability

4. Provide leadership opportunities in order to encourage career development and succession planning.
   - Does not always assume you know people the best
   - Places people in positions where they can demonstrate their potential

- Starts process early in your folks' careers by giving them stretch assignments that provide an opportunity for growth and confidence building
- Plans for small incremental opportunities to prepare people for bigger future roles

5. Draw upon the insights of stakeholders in order to make well-informed decisions.
   - Provides everyone in your organization the opportunity to speak and be heard
   - Helps distinguish between perception and reality
   - Gathers input from people at all levels of the organization to get you to the root cause of issues
   - Helps you make informed decisions
   - Seeks diversity of thought and open communication to help manage expectations when change is inevitable

6. Foster a culture of continuous improvement in order to be adaptive and responsive to a changing environment.
   - Provides positive momentum and keeps employees engaged in a change orientation
   - Creates training opportunities
   - Encourages risk-taking and promotes a mentality for mistakes being acceptable if you are learning from them
   - Encourages employees to look outside the organization for innovative solutions to difficult problems

7. Encourage creative thought in order to promote innovative and effective solutions.
   - Creates cost-saving potential
   - Uses resources efficiently (materials and personnel)
   - Promotes employee buy-in
   - Can result in a safer work environment
   - Encourages communication and interaction

8. Promote teamwork in order to create a collaborative work environment.
   - Promotes employee buy-in
   - Improves morale
   - Improves effectiveness
   - Makes work more enjoyable
   - Promotes mutual respect
   - Breaks down silos and isolation

9. Acknowledge the good work of employees in order to show appreciation for their value and contribution to the services we provide.
    - Documents actions
    - Provides personal communications
    - Increases morale
    - Creates employee loyalty
    - Keeps employees engaged
    - Creates higher performance

10. Hold each other accountable in order to assure personal ownership in organizational performance.
    - Provides for open and honest communication
    - Provides a platform for articulating team expectations
    - Helps define culture
    - Declares that as individuals our choices matter to team performance

11. Use diverse, open, and interactive communication methods in order to promote employee engagement.
    - Helps employees feel more a part of the action (ownership)
    - Creates participatory environment
    - Encourages inclusive ideas
    - Recognizes that different people have different learning styles

Part II

Personal Leadership: Getting the Ingredients Right

The most effective and enduring leaders demonstrate consistent personal leadership through patterned behavior over long periods of time.

5

Values

Success in organizations, and with the leaders who lead them, has a foundation that is built on consistent and defined values that help promote consistent decision-making and behavior.

Key Points

- In the absence of policy, values form the filter by which decisions are made.
- Personal values are the sum of who we are and therefore dictate our behavior.
- Leaders and followers each expect the other to have good values.
- Organizational values are important so that people don't use their personal values to make organizational decisions.

I would like to take a step back from the general topic of leadership and concentrate on its many elements. Think of it like building a house. You don't start building a house by constructing the roof first. Instead, you start with a foundation that is square, plumb, level, and able to support the components of the roof that will reside above. In the case of leadership, our foundation begins and ends with our personal and organizational values.

How is it that two people can read the same books or go to the same leadership class and then one can apply the concepts and the other flounders? We know it's not about being smart because there are lots of smart people who cannot lead. We also know it's not about tenacity because there are many tenacious leaders who basically suck at being the boss. If it isn't about being smart, tenacious, competitive, well-liked, or any other adjective, what in the world could it be?

Could it be values? Could it be those values, loosely defined as the basis for how we think about life and decide how we are going to act each day, that are the answer? Could it be the values that make up those intangible elements of our consciousness that we have decided will guide our actions and decisions? Could it be those principles that we care about most and in many cases admire the most in other people?

When you describe someone as a good person, what are you describing? When you describe the person, is it based on consistency of behavior or their decisions on just a most-of-the-time basis? If a person acts in a certain manner most of the time but slips up occasionally, does that make them a bad person? Have the people you are describing ever done anything that would be inconsistent with the attributes of a good person?

These are all good questions. We all make missteps when it comes to applying our personal and professional values in various situations, but good leaders do not make the career-ending types of mistakes that employees and employers are reluctant to forgive.

What one can say about values, and in turn about someone who lives their life by a good set of values, is that under most circumstances they provide a lens through which the world is viewed, a decision filter of sorts for managing the push and pull of everyday life. Values, in this example, provide a basic set of rules that are applied in every situation so that decisions about that situation are consistent regardless of the varying circumstances.

> Values, in this example, provide a basic set of rules that are applied in every situation so that decisions about that situation are consistent regardless of the varying circumstances.

One consistent factor in long-lasting leadership is that effective leaders have a good value system because it is this value system that helps them make decisions in the absence of clear right or wrong. As the top decision-maker in any organization will tell you, the right and wrong answers are all gone by the time the issue reaches their desk. In these situations, there is no policy that will help because, if that were the case, some lower-level manager would have already solved the problem.

Having good core values is always required for a leader; however, the emphasis the leader places on certain values, applied at the right moment in time, is what defines great leadership. This sense of timing defines the difference between good leadership and great leadership.

When a person says that their most important and guiding value is honesty, it doesn't mean that they will always be honest. Likewise, when someone says they are compassionate, they don't always truly understand every situation that calls for compassion and therefore won't always appear compassionate.

At the end of the day, all humans are conflicted when it comes to right and wrong, and depending on the internal and external pressures of any given situation, anyone can give into the temptation of compromising even their most closely held value. Most of us know that when it comes to the whole right and wrong thing, only one guy got it right. The rest of us struggle daily.

Personal Values

These days, it seems one leader after another has lost touch with a decent value system. From news reports of embezzlement, fraud, misuse of funds, or other crimes, some leaders just don't have a sturdy platform of personal and professional values. Unfortunately, as some people achieve higher leadership roles in an organization, they somehow also seem to gain a proportional sense of entitlement. They simply do not believe that the rules apply to them any longer.

Our personal values are an inside thing. Just like your body needs a proper diet and exercise, your values need constant attention. Just as with personal health, what happens on the outside, with respect to personal behavior, is a direct reflection of what we're made of on the inside. The thing about values is that you cannot just bring them out occasionally and expect them to perform at a high level. They always need to be on the forefront of your thoughts to work in a consistent manner.

Values, and in particular values-based leadership, is grounded in the fundamental premise that if you hire people with a good value system, and the values they have are the same values the organization uses, then you can not only expect an outcome of good decisions, assuming your organization has good organizational values, but even more importantly, the decisions your employees make will be aligned with other employees' decisions in the same organization.

Using values as a foundation, and more specifically our personal values, it is critical that we understand how values affect our decision-making and, even more importantly, how using inconsistent values, shifting values, or allowing the weights of the values to change by the situation places you in a position of leadership inconsistency.

An example would be when you are using personal honesty as a value to guide the organization and one of your employees has been less than honest. Depending on how important this value is to you, it will be assigned a theoretical weight. Some people might hold honesty as a number one value and always expect people to be honest with them. In their mind, there can be no compromise for being dishonest, regardless of the situation's circumstances.

In the case of a police officer, honesty is very important because judges and citizens need to be able to always be assured the testimony of the officer is truthful. It would never be acceptable for police officers to be anything less than truthful in every situation regardless of their personal values.

But what happens when a fire chief decides not to put a great personal weight on honesty even if they work in an organization that places a great weight on the organizational value of honesty?

Let's for a minute say that you are the fire chief and are confronted with an employee who has been less than truthful concerning an expenditure incurred at a professional trade conference. The problem for you is that the employee just happens to be your deputy chief, and to make matters worse, he is one of your closest confidants and friends.

Would you treat your second-in-command the same as a firefighter? Would you treat your closest confidant the same way you would treat someone separated by many levels from your leadership team?

The answers to these questions and many more like them are very difficult personal decisions that leaders wrestle with each day. We are all challenged in these types of situations and can see firsthand how shifting the weight of the value can have an effect of how it's applied from situation to situation.

If honesty is a highly weighted organizational value, then it should apply to everyone in the organization, and that includes your peers in leadership. Many leaders make the mistake of shifting the weights of the values because the person in question is their friend. Not holding one person just as accountable as the next puts you on the slippery slope of poor integrity.

Our behavior, and the behavior of our employees, is a byproduct of the values we use to guide our lives and make decisions. These values then instinctively become manifested in any number of traits that are finally exhibited on the exterior in plain view for all to see and judge.

I have often used the following very effective employee and leadership values exercise to examine the expectations we have of one another when considering personal values. Very simply, divide any group of people in half and have one group take the role of follower and the other take the role of leader. When everyone has a role to play, ask them individually to answer the following question:

- As leaders, what kinds of things do we want from our followers?
- As followers, what kinds of things do we want from our leaders?

When they have completed the exercise, start with either one of the groups writing the results on a flip chart. Invariably, the results will be a combination of traits and values and will most likely include words like honesty, integrity, compassion, trust, technical competence, listening skills, focus, discipline, tenacity, communication skills, and common sense.

Then move onto the other group and hear their answers. They will start with honesty, integrity, compassion, trust, technical competence, listening skills, focus, discipline, communication skills, and common sense. I guess by now you get the picture.

The lesson here is powerful. What leadership expects from employees is consistent with what employees expect from their leaders. In most cases, the exercise will identify integrity, honesty, compassion, and trust as the top four values.

The lesson for the group doing the exercise is that all of us, without respect for position in the organization, expect the same from each other. Leaders expect their employees to be honest, and followers expect their leaders to be honest.

Leaders expect their employees to be compassionate when they make mistakes, and employees expect the same.

A similar exercise can also be used to determine organizational values. In this regard, ask all your employees to list what they consider the top five attributes of great leadership in order of what they perceive to be the value or trait's level of importance. For the purposes of the exercise, do not distinguish or differentiate between values and traits or what levels of leadership you define as leadership.

> From a personal point of view, I am a firm believer that the basis for all good leadership is grounded in the four values of honesty, trust, compassion, and integrity.

When you are done gathering the information, note the number of times each value or trait is listed and how it is prioritized in comparison to other values and traits. For the purposes of determining values, at this time, you will need to separate the values from the traits; however, do not discard the trait data, as we will use it later to determine organizational leadership traits.

When you have analyzed the data, you will very easily be able to articulate the organization's top values by level of importance. What you now have is a filter, of sorts, that all the organization's members can and should use to make decisions.

Your results may vary; however, my experience has shown that most organizations will list honesty, integrity, compassion, and trust somewhere in their top values. In addition, I have never done this exercise where honesty was not listed as the top value.

Here are some examples: In one organization, the fourth value was influence. In this case, the department in question was coming out of a period of leadership in which they felt the leader didn't really lead but rather just followed other departments' leadership. In another organization, the fourth value was compassion, which, as it turned out, spoke to many years of overly aggressive discipline and control. Oddly enough, in later years when this same organization did the same exercise, the fourth value was identified as capable, which was the result of a new officer corps that followed a retirement buyout by the department. Finally, another department's fourth value was loyalty, which was a byproduct of the department's experience with publicly ousting a long-term chief. In each of these cases, and many others, the fourth and fifth values identified some underlying feeling of the employees that dealt with the organization's culture at the time of the survey.

I think it's necessary to draw a distinction between a value and a trait and how each will be presented in this text. This is important because many values and traits can be labeled as either one depending on whether they are being described as an internal belief or external behavior. The best way to draw a distinction between the two, even when the same word can be used for both, is to describe a value as something that can't be seen while a trait can be observed as an external action.

The following is a list of the values often identified as those most admired in successful and capable leaders.

Honesty

Honesty is linked so closely with integrity that the two of them will almost always be spoken in the same breath. In my mind, the act of telling someone something untruthful is to be dishonest, and the overriding morals that allowed the situation in the first place deal with integrity. The reason for that will become clearer in a minute when we discuss integrity, but for now let's focus on honesty as a value.

At one point early in my career, our department spent a great deal of time working on job descriptions and promotional processes. We were trying very hard to make sure that what people were being asked to do was, in fact, the expectation for that job. Once this was complete, we put in place a promotional system that we believed would test for the identified skills. That may seem easy, but it takes a great deal of communication and compromise to pull something like this off. The leader needs to be very clear with folks concerning the expectations of various jobs and make sure it is written down because any subsequent change in promotional practices will cause some people to view themselves as losers in the new process.

The bottom line is that we had worked very hard setting up promotional policies that evaluated the various aspects of the job expectations in which the candidates were competing. The processes had various sections, and all the sections were weighted to 100%. To make things a little more complicated, a few of the sections were composed of individual elements. An example would be Phase One equaling 30% of the final score; however, the Phase One consisted of the written test score at 40% and a personal assessment report (PAR) at 60%. In this situation, the written test and PAR together made up the 100% of the 30% allocated to Phase One.

What I haven't described for you here is the remaining phases and individual elements of the process, but I think you can get the sense for how complicated it was. When the promotional process was over, there were many mathematical inputs that, when complete, separated some candidates by only hundredths of a point.

After the process was complete and the eligibility list was posted, one of my assistant chiefs brought to my attention that there was an incorrect calculation in the first part of the process. The effect of the error had to do with the weight given for various educational levels.

All I could say was "crap." How many people are affected by this? Will I have to put all 60 or so people back through another process? If names change on the list, will the candidates who were already certified in the process file grievances? Worse yet, was someone eliminated from competing in all phases because they were cut early in the process? To say the least, my heart felt like it would beat right out of my chest.

You could pause now and take a few minutes to run through the scenario to decide how you would handle the situation. For my part, even though it was only another chief and me who knew about this, there was only one thing to do, and whether I lost credibility because the process was flawed was not a consideration. I had to be honest.

Egg on the face and all, I told everyone that we had found an error and that based on that error some results would change. It wasn't a pleasant few weeks after that as I had to listen to the naysayer's rumor that the list changed because "my guys," whomever they were, did not make the final list.

The one good thing about admitting mistakes and not covering things up by being dishonest is that while people might think you are a knucklehead, they will at least come to know you as an honest knucklehead.

If you want to be a chief, I will tell you that you only need to be caught being dishonest one time and then you will no longer be chief. Instead, you will be a liar who is called chief. You will be someone who cannot be trusted and who is not viewed as loyal to the service and code. Do not be disillusioned here; people will still follow your directions. However, it will be because of the authority your position holds and not the credibility of the man or woman who holds the position. The stakes are high if you want to be chief, which is why chiefs who are not grounded in good values don't last very long in the position.

Integrity

I am not a sociologist of anything but rather just a fire chief who has taken the time to write down his feelings about leadership. With that as a disclaimer, I have always had a hard time classifying integrity as a value. I see integrity as the bridge that links a person's internal values with the behaviors that ultimately manifest as traits. That said, integrity is frequently counted as the number one characteristic or attribute necessary for successful leadership.

> I see integrity as the bridge that links a person's internal values with the behaviors that ultimately manifest as traits.

I am reminded of a guy standing in line at a grocery store. He was preparing to pay for his items when he spotted a $20 bill lying on the ground. After a quick glance to see if anyone else noticed, he placed his foot over the greenback and then waited for an inconspicuous opportunity to reach down and retrieve his newfound wealth. After some additional assessment of his surroundings, he placed the money in his pocket and walked toward his car.

When I was in Lynchburg, we defined integrity as "doing the right thing when no one is looking." This definition is born of the fact that it's sometimes much easier to do the right thing when everyone is watching your every move, and it's sometimes more challenging for us when we are alone and left to our own devices. That said, I'm sure all of us would have handled the situation described here by retrieving the money and turning it over store management. Or would you? Only you know for sure.

If your efforts at being a leader are devoted to creating a model organization, you will undoubtedly have a lot of work to do. Some issues will be fixed under the watchful scrutiny of organizational leadership and the citizens, and some others will be mostly invisible to them. The question you need to ask yourself remains, are the problems in plain view as important as those that are not? Will you be looking around to see if anyone is watching before you address issues that are not in plain view or perhaps may not benefit the organization's personnel?

As the chief in Wilmington, I was presented with an issue regarding the use of holiday time. The situation was that firefighters wanted to bank their holiday as opposed to being paid for them as the holiday occurred. Like most fire organizations, we worked a 56-hour week, and for conversion purposes a firefighter's day is 11.25 hours. In other words, their 8-hour day is equivalent to 11.25. Since regular employees get ten 8-hour holidays, the firefighters get ten 11.25-hour holidays.

Since I had just started this job, the request to be able to bank holidays seemed like a real low-hanging piece of fruit. New guy goes in and starts by helping the firefighters get better benefits. Slam dunk, right?

As it turns out, the permission to allow them to bank holidays was easily worked out between human resources, finance, and the city manager. In doing the completed staff work (see chapter 22), I happened to uncover that the existing earning rate for leave was not based on a factor of 2,916 hours, which would have created the 11.25 calculation. Rather, it was based on a higher number of hours, and, in fact, their current rate converted to a factor of 11.33.

Now, the difference between .25 and .33 might not seem like very much, but that doesn't mean it's unimportant. In fact, I have found that the two most controversial issues in firefighter-land are leave and pay.

Filed under the heading "no good deed goes unpunished," I fixed the problem of banking holiday time, and the net effect was a reduction in the hours the firefighters earned. So, what is a fire chief to do?

I was the only one who had found the error, although I'm quite sure one day, in the future, some bean counter in finance would have uncovered it as well. Nobody was looking over my shoulder when the error was discovered; however, in my mind there was only one thing to do, and that was to tell human resources and finance about the error.

That did not sit well with my folks, but, in a glass half-full kind of way, it afforded me the opportunity to explain the situation based on my values. In effect, it was a real-life example of how values can be used in making organizational decisions. We did the right thing even when nobody was looking because, bottom line, it was the right thing to do. That is integrity.

Trust

As the chief, you will be miles away from the operational aspect of your service delivery system. Whether it's because you are in a meeting or at home sleeping, it will become very clear very fast that your ability to control what happens on the frontline level is limited at best.

To address this situation, leaders write policies and procedures to govern employee actions. Some departments have volumes of rules and regulations that are controlling in nature and only create what can best be described as a parent/child relationship. That may be okay when you're dealing with parents and children, but it has a negative impact when dealing with adults in the workplace.

If you have ever worked in a parent/child type system, you know full well how suffocating the work environment can be. Leaders that try to lead in these types of systems will fail because, first, you can't articulate a rule or regulation for each and every circumstance and, secondly, your employees will be in a fearful state all the time and will wait to be told what to do. That is not how to run a high performance organization.

Many leaders assume rules and regulations establish some sort of basis for accountability. Their thinking is that, if I can write a rule, I can control behavior, and if I can control behavior, those who choose not to abide by the rule can be disciplined. You know what? They're right, but at what cost? A group of paralyzed employees waiting to be told what to do? A workforce full of people not helping to make things better because they view themselves as being just a whack away from having a knot on their head?

Think back to a time when you were really engaged in helping your organization move forward, and I will bet it was because someone told you what they were trying to achieve and then left you alone to get it done. They gave you the tools and let you get to work. They, in effect, trusted you to do a good job, the key word being *trust*.

> Deal with the folks who get sideways and respect the ones who do not by resisting the notion that a new rule is in order because some knucklehead came to work with green hair.

For my part, I choose to trust. I choose to stand up in front of my folks and say that I will provide you the tools you need to do your job, the vision for what it is we are trying to achieve, and then I will get out of your way and let you get me there. Like baseball coach Casey Stengel once said when someone asked him how he contributed to his team's success, "I don't trip them when they leave the dugout."

Sure, some folks are going to let you down. Not everyone is going to do the right thing all the time, but accountability comes naturally when people have responsibility for their actions and they are not wrapped around the axle of some policy or regulation. Deal with the folks who get sideways and respect the ones who do not by resisting the notion that a new rule is in order because some knucklehead came to work with green hair.

The other aspect of trust is the trust your folks give you as the leader. It's a two-way street, this trust thing, and you will quickly find that, as the leader, it's very easy to get lost. You realize full well that your folks are going to let you down on occasion. Still, you will trust that most will do the right thing most all the time. It may not seem fair, but the point is that there are many of them and only one of you. The spotlight is on you because you are the tip of the spear. You must always do the right thing, and if you can't stomach that kind of accountability, then the top job may not be the best for you.

Trust is developed when people see vulnerability and decide not to take advantage but rather to help despite the vulnerability. Trust is developed when someone truly helps someone else for a reason other than to make themselves look good. It becomes not about you looking good but rather helping someone else look better.

In healthy teams, individuals can sense when someone is in a vulnerable position and then will help that person and not harm them. If your organization is set up based on the achievement of individual goals and not team goals, the opposite will occur. When individual goals trump team achievements, human nature will dictate a survival-of-the-fittest mentality. In effect, when someone is down, the system works to eliminate what is perceived to be a weakness in much the same manner as an injured elk can fall prey to a lion on the hunt.

> Trust is developed when people see vulnerability and decide not to take advantage but rather to help despite the vulnerability.

Another trust killer is when the actions and behaviors of the leader do not align with organizational values. Seeing yourself as the leader who is above your organization's values is a sure way to lose followership.

I have unfortunately worked for and with people who do not value those who are willing to share their thoughts and feelings. As both a follower and a leader, you need to be on the lookout for people who do not appreciate honesty, sincerity, or openness. It is these folks' mission to figure out an angle to use this type of sharing as a wedge that demonstrates to the organization that you have a weakness.

As the leader of any unit, division, group, or organization, you must aggressively seek out situations in which employee behavior is not aligned with the teamwork and trustworthy expectations you have for the organization's culture. Ferret these folks out and get rid of them. At the very least, find a way to marginalize their effect on the organization's culture.

What you should take away from all of this is that relationships drive trust. People on a great team need to truly love each other and

> When individual goals trump team achievements human nature will dictate a survival-of-the-fittest mentality.

want to help each other to succeed in their jobs. In doing so, they make the overall team more effective. Successful relationships are cultivated in an environment where people feel safe: safe to say what is on their minds and safe to say they need help in areas where their expertise is not suitable for what needs to be accomplished.

Trust is developed in a relationship when people are compassionate toward each other about inadequacies and deficiencies, reaching down to offer a helpful hand when needed. There should be no ridicule, no power play, no positioning oneself for success if it comes at the expense of others, just plain old teamwork resulting from relationships that have compassion and love as the foundation.

It's the leader's responsibility to manage the culture of the organization and in doing so create a trusting environment. It must be safe to think, talk, and act in a manner that challenges the norm and reveals an individual's true feelings. Good leaders create situations where the curtain can be pulled back because there are healthy and trusting relationships between all employees and people are not scared of what might be standing on the other side.

So why is trust such an important element of leadership? Well, that, my friends, is simple: the follower must be willing to let their guard down to follow the leader's vision. For that reason, it is imperative the follower first trust the leader, and it is equally important the leader help by articulating the vision to the followers. Less trust is required when followers can see the vision and understand where they are going.

One last word on trust: trust is not authority and control, and frankly it can be said that it is the opposite. Authority and control will hammer trust on the head until it is black and blue. Conversely, trust creates initiative on the part of employees. Nothing motivates employees to work in the best interest of the organization more than knowing someone trusts them and implicit in that trust is accountability for the work.

Compassion

Compassion is a peculiar value, as it seems it is weighted one way when applied to employee issues and yet quite another when applied to leadership. Just like everything else that has to do with leadership responsibility, the stakes are much higher for the leader than the follower. The translation here is that employees expect compassion but are not really the compassionate type when their leader makes a mistake. At the end of the day, maybe it's most appropriate that it's this way, as the burden of being the leader is great on just about every level.

The one consistent aspect of compassion, no matter if you are the employee or the supervisor, is that everyone who makes a mistake wants a piece. I don't know about anyone else, but I have never had an employee come into my office and say, "Screw you, just give me whatever punishment you want." In general, people who accept responsibility and those who don't all seem to want their leaders to be compassionate.

In this regard, compassionate does not mean easygoing, but rather it is a value that speaks to the frailty of human behavior. This aspect of leadership first requires understanding that everyone makes mistakes and, in doing so, we declare ourselves to be human.

As leaders, we have many unfortunate tasks, not least of which is to discipline employees. In the case of firing people, you not only punish the offender but also their family because you have taken away a measure of their financial security and affected their work record in a permanent way. While demotions, suspensions, and reprimands are somewhat less anxiety-producing and emotional, they are still significant events in people lives.

Compassion should also not be confused with forgiveness. Forgiveness is a very personal act, whereas compassion for a situation is a personal value that can be applied to organizational problems. You can choose not to forgive someone for how they have treated you personally—although I have never found this to be a good practice—while at the same time being compassionate regarding the circumstances that led to the event.

No matter if you are a young firefighter just considering leadership or a seasoned and experienced officer, what all of us have in common is that we work for someone else. In instances where I needed to ask my boss for compassion, it has almost always come from moments where I neglected to ask for permission. In each of these circumstances, the one thing I did not do is make excuses for myself. In that regard, I always make a point to just say this is what happened, these are the circumstances, and I accept the responsibility for the results. Manning up, doing the right thing, or throwing yourself on the sword—any of these will work, just pick which one you want.

Early in my career, I accepted an alcoholic beverage from a student at the end of a very long and intense week of training. I knew full well that alcohol was not supposed to be on the training center property. I also knew breaking that rule was a very significant rule violation that could lead to termination. With hindsight, I should have just said no and enforced the rule myself as a member of the organization. It just looked so cold and refreshing. Okay, wait a minute, that is an excuse, right?

The next Monday morning, I was confronted by my battalion chief, who very calmly advised me that he was opening an investigation and that, as per the Firefighter Bill of Rights, I should know that it could lead to my dismissal or suspension. I remember very clearly telling him there was no need to do an investigation, I did in fact accept a cold alcoholic drink from one of the students. In addition, I was sorry for my transgression and understood the violation could reflect on his leadership, and I was very sorry it ever occurred. I rambled a few more sentences around my resolve to step up my behavior and how I was seeking any compassion he might have for a dumbass captain who, at the time, didn't think anything on earth looked better than that cold beverage. Okay, the last part was another excuse.

In the end, I was disciplined, and I was not happy about the situation. It was not the eight hours off without pay that got my goat, heck, I went golfing. It was that I let down a battalion chief and department that was expecting something different from their officers. In the end, I did receive compassion for my mistake, and, to this day, I remain thankful that I reported to a chief who had compassion as a personal value. I must say that it was some time before I forgave the officer who reported me, but maturity and time have a way of making that happen as well.

On the reverse, I have been on the listening end of a litany of excuses for one thing or another. Whether it was "I didn't do it" or a "that person made me do it" or even "the person who saw me do that didn't really see me do what he is reporting," I have heard it all. Even in situations where employees refuse to accept responsibility for their personal actions, I still try to be compassionate. It was how I was treated by a great fire officer.

Getting to Your Personal Values

Now that you can distinguish between organizational values and personal values and understand how and when the two types should be used in decision-making, you need to establish how and why they are different. You can do this from a personal point of view by asking yourself what kind of behaviors are acceptable when examining the expectations you have for your personal life.

Another way to ask the same question is to try to think about what your expectations are for others with whom you share a relationship. What kinds of behaviors would you absolutely consider a dealbreaker in a relationship, and why do you feel that way? The answers to these questions should form the basis for how you will and can do everything from pick friends, to find a partner with whom to share your life, to outline the expectations you have for your children. In each of these circumstances, you are saying to the world that these are the values you hold dear and the platform from which you operate your life. To establish these values and then not use them to guide your behaviors and relationships can create a situation in which some unscrupulous people will have access to your personal situation.

Let's look at a real-life example: You have just been hired to work for an exciting start-up company and the possibilities for your future appear to be limitless. Because you are very talented and professional and have a great work ethic, you rise quickly to the management team. You soon find that the issues on the management team are much different from the issues mid-management deals with, but you are excited about the opportunity to learn and add value to the company's leadership.

During one of your management meetings, the company is discussing how to create additional sales. During an ensuing debate, it becomes clear that the management team has already instituted bait-and-switch sales tactics. In effect, the salespeople offer opportunities that appeal to many potential clients; however, in reality only a very few will ultimately be eligible.

Over the next few weeks, you become increasingly concerned because some of what management told you when you were in your previous position was much different than the types of things you are now hearing. You become frustrated by the direction and leadership. In one meeting, you mention that the message being sent to the employees may be inconsistent with what is being considered as future direction. The boss then makes some statement concerning employee entitlement and tells everyone to just stick with the message.

Over the next few months, you find yourself increasingly at odds with the organization's apparent values and those you have established for your own life. Your personal value of honesty is being cheapened each time you are instructed to act in a manner inconsistent with your personal value system.

As you look at each of these situations individually, you determine that the leadership of the organization has different values that are not aligned with your personal value system. Although you were looking forward to an exciting new career, you determine that this organization is not one in which you can work, and for that reason you resign.

These types of situations play out every day in organizations that are rudderless when it comes to values. The question remains, how could this situation have been avoided? First, the organization could have articulated their values. Without this, how would you ever know that the actions of employees were aligned with your personal beliefs? Secondly, you could have discussed your personal values with your boss, and if they were substantially different from those of your boss and the organization, then perhaps you would not have been hired in the first place.

This situation is no different from picking friends. There are times in your life when you will need to distance yourself from people because their value system isn't aligned with yours. If you aspire to be a leader one day, you must understand that you will be judged in some manner or respect by the people you allow to be in your inner circle. If folks in that inner circle do not share your value system, then you need to distance yourself from those relationships.

The exercise of determining your values is an important tool for your success as a leader. If you elect to put down something that you think is the right thing to put down but, in your heart, you know you don't behave consistently with that value, then you can stop reading right now and just close this book.

As you complete this book, you will be presented with several exercises that, when complete, will leave you with a personal plan for success. Included in this plan are your values, mission, vision, primary focus areas, objectives, and something I call enablers and restrainers. The values part is the first and perhaps the most important activity in your plan because they represent the platform upon which the rest of the plan will be built. Said another way, this exercise is the foundation of a building, and it must be built square and sturdy for the other parts of the structure to fit properly and ultimately withstand the elements of organizational storms.

Understanding your personal values will help you distinguish between your personal feelings regarding a situation and the expectations your organization has regarding the same situation. The bottom line is that great leaders are great individuals who are successful because their life is driven by a consistent set of principles and values from which they will never waiver.

So, what are your personal values? The most effective way to get to your personal values is to ask yourself a series of questions. These questions are designed to get to the root of what drives your way of thinking and, in most cases, the situational responses that result from that thinking.

Write down your answers to the following questions. After you have completed the work, look deep into the responses for the true meaning of what you wrote down. Don't just read what you wrote down but rather ask yourself the meaning of it, as that is where your true personal values will reveal themselves.

- What do you admire the most in others?
- When you make decisions concerning your actions, what do you use as behavioral boundaries?
- When it comes to dealing with a situation in which choices of action are required, what will you always refuse to do?
- What do you least admire in others?

6

Traits

Great leaders develop and nurture quality personal traits that can be patterned over time in order to demonstrate consistency in leadership behavior.

Key Points

- Traits can be described as an external manifestation of behavior.
- Traits can also be values depending on the application and context.
- Effective leaders use traits in combinations that help the organization.
- High performance leaders exhibit dominant and consistent traits.
- None of us are good at all traits.
- Traits can be varied in scope and application to be effective.

The discussion of values was presented first because, as I mentioned, it's the foundation on which everything else is built. The resulting behaviors that promulgate from values are presented as traits. Great personal traits do not come from poorly grounded values.

Merriam-Webster defines a trait as "a distinguishing quality (as of personal character)." A trait is something we exhibit as an external quality, while a value is an internal characteristic. Traits and values work very closely together because internal values determine distinguishing traits that are exhibited in how we behave.

The example I like to use is that integrity, which is a value, is the cornerstone of credibility and various other traits. Positive credibility depends on good integrity in as much as you will have no credibility if you have no integrity. People will not see your integrity but rather how it is manifested as credibility. They will watch how you conduct yourself over time and then draw a conclusion about your character based on the traits you exhibit.

While it may take a while for someone to draw a conclusion regarding your values, they will notice your traits right away. If you do what you say you are going to do, you are considered credible. If you display confidence in the face of adversity, you are considered confident. If you demonstrate patience when folks need time to resolve issues, you are considered patient. Here are a few questions you can ask yourself for reflection: Are you on time for meetings? Are you disciplined to maintain your wellness and fitness? Are you considerate of other people's feelings? Are you a good listener? These are just a few ways people define our traits and therefore draw conclusions about our character.

> The lesson here is not so much that people have different traits but that effective leaders use traits in combinations that are effective.

One aspect of traits that all of us need to be aware of is that we are not going to be good at all of them all the time. In much the same way as you would want to leverage your personal strengths in a positive manner, understanding personal traits that are a weakness can help you minimize any potential negative impact they could have in your life.

A good application would be to create a management team where the traits of the members are stronger in areas where you need the most help. If you have difficulty empathizing over employee personnel matters, surround yourself with people who have empathy as a strong trait. They can advise you on how employees perceive a situation, and therefore your subsequent actions won't come off as cold and uncaring.

Think about traits as a bag of tools. When the appropriate time calls for a certain response, the effective leader just reaches in the traits tool bag and pulls out the required tool.

Personalities of all types have demonstrated leadership over the course of time, according to an article by Jerry Useen in the November 12, 2001, issue of *Fortune* magazine. The article examined whether a correlation exists between a leader's personality and their effectiveness and concluded that the meek and mild, as well as the flamboyant, have been equally effective as leaders.

So, if leadership is not correlated to personality type, what determines a leader's effectiveness? Could it be the individual use of the traits that make up the personality type? Perhaps it is the timing with which effective leaders use the traits that make up their personality? The lesson here is not so much that people have different traits but that effective leaders use traits in combinations that are effective. In this way, the leader's actions are matched correctly to the event and the outcome that is most desired.

While the decisions of leaders are driven by their personal value system, the way the actions manifest themselves at any given time is more closely aligned with their personal traits. The list of traits that follows is not exhaustive, but rather includes those that I have found are exhibited by people considered great leaders and relationship builders.

Confident Humility

Being humble and appreciative that your followers are willing to follow you is a trait not easily taught. Humility is the self-deprecating aspect of human behavior that says to people, this guy puts his pants on the same way I do. Humility also allows the leader to let their guard down and say to the organization that some things need improvement. I frequently remind folks who work for me that I truly see myself as only one aspect of what makes the organization run, and even in that role, I am not good at a lot of things. Letting people know your strengths and your weaknesses is what humility in leadership is all about.

Early in my career, I spent a great deal of time training in special operations. The absolute icon of special operations in the American Fire Service was Ray Downey, chief of special operations in the New York City Fire Department (FDNY), and one of the most decorated firefighters to ever wear an FDNY uniform.

At this point, you must put in perspective the distance between Ray's importance to the fire service and

> Humility is the self-deprecating aspect of human behavior that says to people, this guy puts his pants on the same way as I do.

some dude in Virginia Beach tying knots on the top of a rope tower. He was a published author and was referred to very affectionately as "God" by his own personnel. By all accounts, I was still wet behind my ears.

The first time I met Ray, I was invited to help develop one of the training modules for the Federal Emergency Management Agency (FEMA) Urban Search and Rescue (US&R) Specialist program. Ray did not arrive with an entourage and turned out to be a down-to-earth, very humble man, well known for how well he did his job.

As I reflect on my friendship with Ray, I always seem to define his greatest trait as being humble. He was humble that he had achieved great things in the fire service but, at the same time, mindful that it was others who helped him get there. He was humble enough to sit and talk with the newest recruit and then get up and have the confidence to work side by side with the most powerful politicians in Washington.

Our FEMA team's first US&R operational deployment was to the Murrah Building bombing in Oklahoma City. I don't really remember the walk to the scene; my mouth was so dry I'm not sure I could have mustered up a good spit. We approached from the opposite side from where the bomb went off, and when we rounded the corner, it was some sight: 11 floors of building reduced to a 35 ft. high pile of rubble.

As I stood in front of the twisted metal, crushed beams, broken glass, and other related office products, it was hard to concentrate on the enormity of the situation. A leg sticking out here and an arm there, I found myself almost transfixed on the very real and, at the same time, surreal scene. I will never forget seeing the red stripe on the blue pants of a Marine Corps recruiting officer, trapped under a desk and a very large column. It would be several days before we could free him; the daily sight was a horrible reminder of the price these innocent people paid.

During my first few minutes of taking all of this in, Ray approached me. "We got a heck of a mess here. People are still trapped and there is an active amputation extrication going on in the basement. I think I would like to get you guys working on the front of this pile. What do you think?"

What do I think? At this point I just needed and wanted some direction. It is this leadership trait that defined Ray Downey. He was extremely competent and confident but never afraid to say that he did not know what to do next. He was good at his job but also humble enough to ask for help. It was this combination of confidence and humility that made Ray such a great leader (fig. 6–1).

Ray, along with many other very brave firefighters, many of whom were my friends, died on 9/11. It is their legacy, the kind that Ray left by virtue of how he lived his life and treated other people, that continues to live on. Ray's greatest trait may have been humility, and that, my friends, is what humble looks like.

Credibility: Hard to Get and Easy to Lose

If integrity is defined as doing the right thing when no one is looking, then credibility should be defined as consistently doing the right thing for the right reason. It's very important not to minimize how important consistency is to credibility. Consistency of action, when dealing with similar situations, is what builds

FIGURE 6–1. Fire Chief Ray Downey with the author after the Murrah Building explosion in Oklahoma City (courtesy of Bob Anderson)

credibility and leads people to characterize someone as credible. For instance, are you always honest with people even when it could be embarrassing and counterproductive to progress?

The one thing that all leaders find out very quickly is that it's difficult to lead without credibility. Your people may, in fact, follow you for a time based on trust and respect for your position or your background; however, credibility is required to successfully solve deep-rooted organizational and cultural issues.

> Consistency of action, when dealing with similar situations, is what builds credibility and leads people to characterize someone as credible.

Since credibility is based on the perception of the follower, it can either be earned over a long period of time or be achieved based upon the successes of a single incident. Credibility can also transfer from place to place if the leader has been successful in each of their previous jobs. For instance, a football coach with a lifetime winning record finds that credibility will follow them from one coaching job to another.

New leaders need to especially be aware of this because credibility as a worker does not transfer to credibility as a leader. That is a big mistake people make when they're first promoted. Credibility established by being a firefighting technician does little to impress folks when you are responsible for leading them as an officer. First-time supervisors need to be aware that their new position is not just technical but rather requires both technical and leadership skills. Their job is just the first step in a first-line supervisor being evaluated for credible leadership.

In organizations, it's easy to spot those not being credible. Believe it or not, there may at times be those working among you who don't have the collective best interest of the department in mind. Even as you may operate an open-and-honest organization, some folks will take issues and mold, twist, or otherwise contort them to have meanings other than those that were intended.

The thing about credibility is that it's much harder to gain compared with how easy it is to lose. It's also garnered over time with little deposits that, like building blocks stacked on one another, get higher and higher. When the stack is high enough, a thing called trust develops. Once you have made enough deposits and your stack is high enough, people start believing what you say and trust that you will conduct yourself in a manner that is helpful to them both personally and professionally. This happens even when not everyone has all the facts in a situation; they just trust.

We are not talking about agreement in this situation because there are times when I don't agree with what my bosses are doing or the direction in which they are taking the organization. This is natural. When these situations occur, it turns out that I almost never have a full appreciation for their point of view or the external pressures that determined the course of action.

The key to credibility is the action that takes place prior to receiving the full disclosure of information. This is the time when you find yourself in disagreement but are devoid of details and context. In these situations, do you speculate in a manner that comes across in a conversation as fact but is not? Do you deliberately mislead people to discredit someone else or the organization? Do you paint yourself as the potential savior of the situation in case it turns out that the decision was in fact not a good one? Are you the type who suspects something is wrong but waits for the organization to make a mistake before engaging in proactive discussions? Do you participate in perpetuating rumors that you know full well have very little foundation?

Leaders of all types get involved in these kinds of situations and search for an effective means to handle the folks involved. The first thing you should do is to forgive these folks for their transgressions. I say this because people who do this kind of thing are seeking your approval. They need to have their egos stroked to make them appear credible. That is why they make things up; they want to appear credible. What they don't realize is that, by conducting themselves in this manner, they are bringing discredit to themselves, the folks they work with, the department, and the profession. I also believe that, if given a choice, normal folks do not want to be around this type of person and, in fact, almost assuredly never want them to be their leader.

Because all of us are leaders in our organizations, we all have a responsibility to give and receive accurate information. I have always maintained that I will never be dishonest with any of my employees and will provide them with every fact that I can. In those situations where I can't, based on privacy issues or just plain bad timing, I hope that they will give me the benefit of the doubt. In turn, the benefit of the doubt is what I always give them.

So, as the leaders, keep working hard to make those credibility deposits. Help your peers make credibility deposits. Be committed to building your credibility deposits higher and higher until such time as you become a credible and successful leader.

Fairness

An employee's view of fairness depends on how the situation affects them personally. In general, they view fairness by thinking that if a coworker did something and got a certain response for it, they should also get the same outcome for the same thing. We all know that leadership and discipline aren't that easy. In general, I guess if life were devoid of special circumstances, like speeding tickets, then fair could be defined as doing the same thing under similar circumstances. The problem remains that life and employee discipline isn't the same as the fine for going 55 in a 35 mile per hour zone.

So how do we define fair in the case of personal leadership? Well, if it were easy, everyone could do it, and you would not be reading this book. As the leader, it is imperative that you take all matters into consideration when addressing how to handle a particular situation. That's right, you must figure out what is the right thing to do based on the circumstances you are presented.

This is especially true when it comes to employee discipline because even our best employees make mistakes from time to time. That's why your efforts at employee discipline should consider what will be required in each individual circumstance to cause a change in behavior for that employee. It's different for different people; I'm convinced that there is a difference between a very good employee who makes a mistake and troublemaker who continually makes them.

I had a situation once where an employee got popped on a random drug screen. In this case, he took a prescription pain medication for a back injury that belonged to one of his relatives. In doing so, he thought he was taking the same kind of medication he already had a prescription for. The problem was, he got hydrocodone and oxycodone confused and took a medication that he didn't have a prescription for.

Now mind you, it is still illegal to take a prescription medication that you have no prescription for, although I think even the most ardent rule follower could make a distinction between this scenario and one in which an employee was using a narcotic with no prescription. The problem for me remained that our city had a no-tolerance drug policy: you get popped, and you get fired. No questions and no excuses. So, I had to fire a good employee who just happened to make an error in judgment.

I have also had employees who smoked dope while off duty and got popped by the same policy. A previous department of mine also had an employee who tried to buy cocaine from an undercover officer one evening. That did not turn out to be one of his smartest moves. When the deputy chief went over to the jail to tell him that his services would no longer be needed, the employee was surprised. As it turned out, the employee couldn't understand why he would be fired when the substance was not cocaine but rather just something that looked like cocaine. Just for a second, imagine the look on the deputy chief's face as he communicated with this employee

> As a leader, you should try and create a culture in your organization that tolerates differences in employee discipline based on an evaluation and consideration of the specific circumstances involved.

through a set of bars and the employee couldn't understand the difference between trying to buy a drug and buying the drug. On my part, I just attribute that to a dumbass test, and that guy failed.

The point to all of this is that each situation, no matter how egregious or stupid, has an individual set of circumstances that led to the outcome. As a leader, you should try to create a culture in your organization that tolerates differences in employee discipline based on evaluation and consideration of the specific circumstances involved. This type of culture requires trust between the leader and the followers because, in most cases, the specifics of the situation will be confidential in nature.

The other thing I question is the organizational use of policies that mandate a specific discipline for a specific infraction. When you have a no-tolerance policy on anything, you should question yourself concerning its real-life application. Is the policy itself just in place to make difficult organizational decisions easy for leadership? Is the policy in effect to create the perception of consistency and fairness when in reality it takes away your ability as the leader to create consistency and fairness?

Fairness is and always will be evaluated on how it individually affects someone, and if that's the case, please take the time to consider how individual circumstances may differ from one situation to the other. Your folks might not understand the differences all the time, but as the Michael Jackson song says, "I am looking at the man in the mirror," and that is ultimately who needs to be accountable for the decision. In the end, just be fair. It's what our employees expect and need from their leadership.

Respect

Respect is gained through personal relationships and the modeling of behaviors over time. In much the same way that credibility can come from past actions or current behavior, respect is given based on past actions and real-time observed behavior that is patterned over time. Remember that, regardless of how or why, respect is indeed given to the leader by their followers, and therefore it can never be demanded.

In my experience, respect is one of those traits that is easily recognized by employees. You see the employees closely watch the way you interact with different levels of the department. Do you listen as closely to the recruit as you do to your deputy chief of operations? Do you cut people off as they are talking to you? Do you listen to what people have to say or act disinterested? Do you make it a habit to look across the organization's many levels, or do you tend to look down the chain from your perceived perch?

There is one thing about the fire service that I don't particularly like, and that is the way rank causes some of us to behave. In many cases, chiefs look down the chain of command like they know everything, and firefighters sometimes look up the chain thinking we do. Here is a secret for you: none of us know everything.

When we first enter the fire service as a recruit, we are like a sponge and eager to learn. Just like little kids full of questions, we fully inquire about everything under the sun. Those of you with children know exactly what I mean. You know the endless questions of a young mind that is not yet clouded by an external expectation that some things should already be known. Why does the sun shine? Why do frogs jump? Why do birds fly? Why were you wrestling with Mommy in the bed last night? Okay, maybe only a few of us have been asked the last question.

> I believe that when you do something nice for someone and have an expectation that a reward will come, it takes all the fun and satisfaction out of the whole situation.

The point here is that it seems, as we move upward through the ranks to an officer level, our natural learning instincts become clouded by our growing egos.

We forget that we can still ask questions. It's hard for us to admit we don't know something or that we need help with something because we aren't good at it.

In these situations, our ego is our worst enemy because it fools us into thinking we know enough to do our jobs and can therefore stop concentrating on learning and improving. I would hope that if you're reading this, you already recognize the value of continuous improvement through learning and how that benefits our customers, not to mention how it makes us better people regardless of our rank or time in the fire service.

> The leader who values respect recognizes that everyone in their organization is a leader. As such, they encourage employees of all ranks to "look across" at each other to learn and not up or down.

I'm sure many of you have also worked for those people who put themselves on a pedestal because of rank or experience and who think it would be below them to admit a mistake or to ask someone a question for fear of ridicule or embarrassment. These folks are missing a great deal in life when it comes to how we can benefit from relationships with people from all walks of life when respect is a key component.

> Many seasoned leaders don't try to change too many significant things when they first move to a new assignment or job, and those who try to effect change too quickly are many times unsuccessful.

The leader who values respect recognizes that everyone in their organization is a leader. As such, they encourage employees of all ranks to "look across" at each other to learn and not up or down. Recognize and value the strengths of all your members because it is in these relationships that you will cultivate a great organizational culture.

As a confident leader, you should display respect for all persons in your organization, and when you do you will get

> We need to raise the analytical bar, so to speak, and change the public's perception that the fire service is a blue-collar job and not a professional occupation.

respect and trust in return. You, as the chief, should realize that although your position may be high ranking and important, it is no more important than those last recruits you just hired. Remember, respect is not about your position in the organization, it is about the value you bring to your position.

Adaptability

When I was asked by an officer in the Wilmington Fire Department if we were going to do everything that Virginia Beach and Lynchburg did, my answer was both yes and no. Like steering a boat in a rocky river, yes, we will borrow from those organizations that have been successful at knowing where the rocks are. But also, in other situations, we may find that the path where those rocks have taken us is not the direction we need to go. In that case, we would go underwater and find our own rocks. More on the boat story in a few pages.

Abraham Lincoln said, "The dogmas of the quiet past are inadequate to the stormy present. The occasion is piled high with difficulty, and we must rise with the occasion. As our case is new, so we must think anew and act anew."

To be adaptive, you are proclaiming that to effect change you must institute practices that help employees see, act, and think anew. It is only through new and innovative thinking that positive change will continue to take place in your organization.

It may also be time for many of you to release the institutional practices that have, for so long, been a part of your organization. This includes openness among members to discuss and freely describe their anxiety about the present and your organization's vision for the future. Most importantly, this includes describing, benchmarking, and resolving amongst yourselves what success in the future looks like.

> How innovative your organization is may, in large part, be dependent on how adaptive and open you are as the leader.

Being adaptive is an important trait for today's modern leadership. Let's face it, things are changing at a speed that none of us could even comprehend a few years ago. The speed at which technology is changing and the advances in the way we provide EMS and command firegrounds are astounding and will continue to push the limits of our personnel both intellectually and from an anxiety perspective.

Being adaptive can be described as managing change. How innovative your organization is may, in large part, be dependent on how adaptive and open you are as the leader. Are you stuck somewhere and having trouble moving on to a new way of doing business? If so, you may be in for a tough time regardless of your being the leader or the follower.

Remember when strategic planning was about short-, medium-, and long-term objectives? Guess what? Strategic planning is now something called adaptive planning. Successful organizations are continually evaluating their direction and goals based on the weak signals on the horizon (those things that can barely be seen now but will have a profound effect on your organization in the near term).

The point here is that previous generations of leaders had the luxury of watching change happen at what seemed a snail's pace. In many cases, they could effectively wait the change out if they were having a hard time being adaptive. These days, change is happening so fast that we can barely manage one change effort before another one is placed upon us. Successful leaders manage adaptability by creating an environment of collaborative experimentation. They make it okay to try things, and if those things don't work, they change them again. No harm, no foul.

Albert Einstein said, "We cannot solve the problems of today with the thinking we used to create them." Modern leaders have an opportunity to do something profound by maintaining a commitment to adapt and manage organizational change. They will be the leaders remembered for changing organizations so future employees are both proud of what was accomplished in the past and challenged by how high the bar is set.

Vision

Vision can be characterized as the ability of a leader to describe an organization's future course while establishing appropriate goals and setting priorities that will get to that place. How this occurs is somewhat of a give-and-take between what the organization can tolerate in the way of change and the capacity of your folks to implement the change. Just as with change management, getting people to buy into a vision that they cannot currently see is very much a leadership challenge.

The first time I went to a NASCAR race was with my good friend Chief Steve Cover of the Virginia Beach Fire Department. There I was, side by side with NASCAR fans who, I might add, are the most uninhibited cross section of America that anyone could possibly imagine. After I got over the shock of sitting beside a 300 lb. dude with a sleeveless shirt, number 3 hat, two packs of Marlboros, a headphone and radio set that only a nuclear engineer could operate, and what I am sure was two cases of Budweiser, Steve took me down to the fence line of the track. It was quite something to be only a few feet away from cars that were traveling at 150-plus miles an hour.

As the cars went speeding by, I looked straight ahead while aligning my view horizontally to the track, and it occurred to me that if, somehow, I had been placed there and didn't know what I was seeing, it would have been impossible for me to explain the event. All I would have been able to describe is what I could see: shirtless men wearing black number 3 hats, yelling at the top of their lungs with a beer in one hand and, almost in unison, pointing in the direction of the first turn. I would come to find out that the number 3 car was Dale Earnhardt.

I think this is an excellent way to illustrate how difficult it is for leaders to provide effective vision for frontline employees who are so close to the service delivery. In effect, they are the ones standing close to the fence. This is particularly true when evaluating what someone else is trying to describe to you concerning the organization's future—a future, I might add, that may be a very different reality than what many people thought even just a few months ago.

I truly believe that a leader does not take anyone anywhere. What they do is describe a vision of where people need to be so they can take themselves there. Great leaders look on the horizon for emerging trends and then make sure that the direction they are describing matches the expectations of what is required to be successful when they get there.

> Great leaders look on the horizon for emerging trends and then make sure that the direction they are describing matches the expectations of what is required to be successful when they get there.

I have a nicely lacquered boat paddle hanging above my office desk. More than a few employees have come in and asked about its backstory. One person told me they thought I might be a boat enthusiast or something. Still another thought it might be an award of sorts. The paddle, and the symbolism represented, remains one of my most cherished possessions, and I will shortly explain why.

My real introduction into leadership began on my second assignment as a battalion chief. Sure, I learned a great deal about leadership as a captain, but I was never tested. As a captain, I had great firefighters because I always worked in the busiest companies. Life was good, so to speak: run fires, train hard, stay fit, get the job done, and be a part of the team.

Leading up to my first day as a leader in a new division, I struggled to come up with a story or metaphor that would help illustrate how important I thought it was to work as a team and drive toward a united vision for what needed to be accomplished. Well, that is where the boat paddle came to life.

Gathered in a conference room, some 50 employees were waiting to see what the new boss was going to say. After brief introductions, I told them our future success was going to be like that of a rowboat. I needed folks on both sides of the boat to row. In addition, I needed them to row in unison with each other because to do otherwise would cause the boat to run in circles. I also needed them to look around as they were rowing to make sure nothing was coming at us from the sides and, furthermore, to look out over the horizon in front of them on occasion.

After some laughing and joking around, I think folks started to recognize the symbolism, and I moved on to what I perceived to be my role in our travels. I mentioned that my main task was to keep the boat (the department) steering toward our goals and to correct the balance in the rowing should one side of the boat start out-rowing the other. I really wanted to be the one who made sure we were headed toward our division and department vision and then make sure we arrived there in the most efficient and effective manner possible.

After more discussion and a few more laughs concerning who would get thrown out of the boat first, and what mutiny on the high seas would look like, somebody ask a very profound question: What is our vision, and where the heck is the boat going? I guess that's the point when I learned that leadership was more than helping people get someplace. Equally as important was helping them see the place we were trying to get to. At that point, we decided that before we started a lot of rowing, we would first decide where we wanted to row. We accomplished that with a strategic plan that started with a vision statement.

Over the years, I have witnessed the power of a vision and strategic plan. I'm convinced that most people in every organization will want to help you get to where you need if you just tell them where that place is located. That is what a vision is designed to accomplish.

Strategic planning and visioning, in the true sense of the discipline, is a big deal. It takes lots of organizational effort to accomplish, and when it's started in earnest there is no turning back. It will consume most of your organizational effort for perhaps a year. It will also include lots of participation from a wide cross section of your department.

It will take leadership to describe vision and help employees see it through to reality. During your journey, you will need to help employees overcome the anxiety associated with change and understand that changes in outcomes only come with changes in actions. It will, at times, be hard for them to step back and see the big picture because stepping back will require great sacrifice. One lesson of life that I have learned is that all things worth achieving are only gained through sacrifice.

The paddle I have on my wall is signed by the members of my staff. It is a constant reminder that people will engage and help you get to your destination when they all understand and see the same vision of the future. Those folks rowed like you would not believe, and I am proud of them to this day (fig. 6–2).

Communication

Great storytellers and great leaders just happen to be great communicators. Ronald Reagan was known as a great communicator because he was straight-

FIGURE 6–2. The author used an analogy about a boat and paddling in unison to reach goals. Fire Inspector Don Moss (right) awarded him a paddle.

forward and honest in his communication. Some would argue that he was too plain and not articulate enough for a president; however, I feel that most would say that when he spoke, they understood. There is a great lesson in that aspect of leadership because if you confuse those that are following you, they will surely get lost.

A main aspect of leadership and being an officer is to filter communications and determine what is important for our folks to know. To that end, you need to adopt a system of communications that, to the highest degree possible, is very open—no secrets. The compromise on this process is that you may go too far and give too much information. That is where the leader needs employees who engage and perform the "receiver" part of the communications process.

Being good at communication is twofold. One, you must create an environment in your organization where accurate and timely information flows both up and down. Because people learn in different ways, you cannot have too many ways to communicate the organization's direction and initiatives. Policies, procedures, guidelines, memos, newsletters, Facebook, Twitter, websites, meeting minutes, video chats, open forums, open door policies, and email all provide

> Every employee should have an expectation that communications of a certain critical nature will be delivered consistently and in the same manner.

a mechanism to get information flowing in your organization. The only caution I would suggest is to make sure people know the difference between routine organizational communications and communications that are critical to the mission and operation. Every employee should have an expectation that communications of a certain critical nature will be delivered consistently and in the same manner.

The second aspect of good organizational communications is to maintain an open and honest organization. For my part, I don't try to bullshit folks because, for one, it's demeaning to the person receiving the feedback, and two, it just isn't credible. I find that people deal with bad news much better if they hear it first-hand and not by the rumor mill, so don't hide things from your employees unless you are prohibited by policy or the political timing is not favorable. In either case, tell them as much.

During some lectures I perform a communications exercise where I have people assume the roles of a probationary firefighter, company officer, battalion chief, deputy chief, and chief. I then assign each of them a corresponding amount of time in service.

Starting with the firefighter, I ask them the kinds of duties they are responsible for in the organization and then repeat the same with each rank. It becomes very clear to most students that if run properly, the various levels of the organization have very different types of duties. In a nutshell, each level has different concerns and, therefore, filters information from the aspect of how whatever is being said will affect their role in the organization's work.

> If you know something is going to be difficult to manage, it is best to get it out in the open where people can effectively talk about it and, if necessary, grieve over it.

In addition to this complexity of communication, each one of the ranks has developed biases over time that are the result of previous experience. Like coffee filters stacked one on top of another, each year of experience builds upon the other until the effect is a reduction in the strength of the coffee, in this case, the quality of the original communications.

As with any change, I recognize that there is anxiety about what the future will hold. My commitment to each of them is that, if the changes we implement do not work, we will change things again and again if that is what is necessary to get it right. What we will not do is be afraid to change because we are fearful of change.

What I have found is that if you know something is going to be difficult to manage, it is best to get it out in the open where people can effectively talk about it and, if necessary, grieve over it. In this way, many of your late adapters will come forward and you will have time to address their concerns prior to actual implementation of the plan.

Being a good communicator also means understanding that the organization is full of different people at different ranks with different life experiences. For example, you have recruits with very few years of experience and senior personnel with years and years of experience. This example of how individuals receive communications has a great deal to do with the context and breadth of their experiences.

To be a successful leader, you need people to act without specific direction to operate the system and make improvements. This process also lends itself to potential gaps in our ability to communicate real-time progress with employees and our customers. Successful leaders find ways to close the communication gap so that individual actions are known to all relevant members of the organization. The actions are further coordinated with other members and aligned with the mission, purpose, and values of the organization.

Optimism

Leaders of organizations can be many things, but as I reflect on my experiences, I think it is safe to say that very few great leaders view themselves as pessimists. You just don't see that many visionary leaders walking around feeling sorry for themselves and claiming that the sky is falling.

Conversely, you do hear about many great leaders who share an overwhelming optimism about life and the work of leading organizations. These folks always think things are going to get better and that all setbacks are a potential lesson to be learned so that they and their organization can become better.

In the fire service we, unfortunately, tend to create officers that stay focused on all the things that are wrong. Like a black cloud of doom and gloom, these officers are forever waiting for the next shoe to drop or organizational catastrophe to reveal itself.

Most of these folks did not get this way by accident. Think about it for a minute: we train them from the time they are hired to be fixers. Having an organization full of fixers is one of the things I both love and hate about the fire service. What I mean is that we are brought up in a system that thrives on recognizing a problem and then immediately reacting to fix it, regardless of the complexity.

It is no mistake that we ended up this way. Think about how you were first trained in firefighting. One day, some veteran firefighter told you that when the bell rings you need to hurry to the truck, put on your gear as fast as possible, quickly drive the apparatus to the scene, quickly size up the situation, decide on an action, and then in a matter of moments affect something positive about the situation. This is our life and what we are paid to do . . . recognize, react, and perform.

This works well on the fireground for several different reasons. The first is that we are almost always reacting and acting with limited data. What information does the dispatcher have? Is it daytime or nighttime? Can I expect people to be home? Will bystanders have any additional information when I arrive? Is the wind blowing? From which direction is the wind blowing? What type of construction is on the property involved? How much of the property is already involved? Do I have exposure problems? Is there an adequate water supply to counteract the level of involvement? Will I be able to access the fire area to deliver the water? Do I have enough help coming?

> In many cases, we take the same fireground problem-solving techniques and try to apply them to organizational problems that are not emergencies but many times much more complex.

As you can see by only this small list of items under consideration while responding to a fire, we do indeed get very comfortable making decisions on the fly and without the benefit of a full set of data. The bottom line is that we do most of our work quickly, frequently flying on nothing but intuition and experience.

One of the great challenges for us as firefighters is to slow down the horses when there is no emergency. You see, it has been my experience that, in many cases, we take the same fireground problem-solving techniques and try to apply them to organizational problems that are not emergencies but many times are much more complex. This process can be fatal as we tend to make decisions based on our perception of the situation and not necessarily on the facts. It is just that we like finding the quickest way to solve problems, and therefore, we get lazy in some cases and just don't put in the time necessary to get it right the first time.

> An effective leader needs to understand that not everyone works fifteen-hour days and, yes, there are people who don't spend every waking hour thinking about the fire department.

> Situational awareness is the ability to understand the environment you are in and then make decisions that correspond correctly to that environment.

The remaining aspect of "firefighter trained behavior" that has the potential to cause us to stumble organizationally is that we are always focused on fixing. We see something broken, and we remain steadfast to resolve the issue as fast as possible, regardless of politics, finances, or full stakeholder input.

Sitting around the kitchen table, we all constantly stay focused on what is broken. It is our nature to look at the world this way and consequently what makes us great problem-solvers. It also makes us seem like "whiners" to people who don't understand our culture. It's not that we don't appreciate problems that are fixed; it is that we find it useless to concentrate on them any longer.

Both situational awareness and leadership timing are for the most part intuitive in that they are not easily taught in a classroom or for that matter even taught as the result of real-life experience.

Okay, so let's face the facts. Bad things are going to happen, and when they do, employees look for leadership to provide clues about how they should feel and behave. Heck, as a fire chief, I bought stolen portable radios for heaven's sake. Talk about a bad day!

The point here is that great leaders do not ignore what is wrong and instead spend lots of time looking for potential problems before they occur. In doing so, they never get to the "sky is falling" and "woe is me" part of the situation. Instead, they stay focused on the accomplishments and positive contributions the department has achieved while many

Most city managers are comfortable seeing life as little steps in one direction and not the big picture with no little steps.

times having to remind our fixers that hundreds of good things have been done.

Still, as I travel from fire station to fire station, the conversation is always about the next thing we need to fix, and that is not always the healthiest way to run an organization. Think about people who only look at the glass half full or only talk about the things that need to be fixed. They are unhappy people who, in most cases, are not just unhappy here in their job but also unhappy within their personal lives as well.

An effective leader gets people in the organization to the place in the organization where their best skills can be most effectively utilized.

Firefighters are great for getting things done, and no leader wants to sacrifice that aspect of our culture. Great leaders also do not want people sitting around thinking about all the things that have not been done yet without equal consideration given to all the things that have. For my part, I choose to be happy, optimistic, and celebrate all the wonderful things my folks do while, at the same time, fixing the stuff we do not do well. The point is to not forget the optimism and celebration part!

Timing

The success of any leader is often dependent on their ability to recognize when the timing is correct to accomplish a given objective. Sometimes this timing involves the politics of the situation while still other times it is dependent on the availability of organizational resources and capacity to accomplish a given objective.

Whatever the circumstance, some leaders get the goal of timing while for others this is a concept as foreign as fourth-year French. The bottom line is that very effective leaders have a keen sense for when it is time to lead.

Leadership timing can be shown in examples that range from the conference room to the fireground. The way leaders handle the timing of issues does, in large part, determine how successful they are at the end of the day and how functional or dysfunctional they are at being involved in the organization's business.

So, what does it mean to have a keen sense of timing? Consider the chief's involvement in a staff meeting where a contentious issue is being discussed. Battalion Chief Jones offers a point to which Battalion Chief Smith offers an opposing view. Your public information officer chimes in and, instead of clarifying the issue, brings up another point that is not necessarily salient to the battalion chiefs' discussion. About that time, the operations chief steps in, and, being a type A "I am in charge" kind of person, states an opinion that gets all others not currently involved in the discussion angry as well. Meanwhile, the chief sits idly by while the entire staff effectively jumps off a cliff.

> The ability to lead is born in self-awareness and confidence gained in one's own experiences. This confidence is gained from leadership experiences where both success and failure are the result.

In this example, there would come a time when a leader who has good leadership timing would recognize the difference between a discussion around discovery of an issue's many facets and the all-out controversy caused between two battalion chiefs who don't agree. At some point, the good leader steps up to clarify the issue, offer a different perspective, or, for that matter, make a decision that allows everyone to move forward. The problem is that many times the leader will sit silently and watch each member of their staff descend into arguments.

In still other cases of missed timing, the chief will stop the conversation before the issue has been fully vetted. In this situation, the members of the group feel that their input is not valued, and they will become silent when you need them to be engaged.

As a leader, you should always ask your key folks to help solve issues. Solving issues or deciding direction in a vacuum devalues your people and leaves them feeling that you lack confidence in their ability to solve tough problems. I can't tell you how many times I have been involved in budget discussions where a senior official told me what to cut in my budget instead of asking me to use my professional judgment and determine which cut would have the least impact on my service delivery efficiency.

The way you, as the leader, behave in these situations is critical, as your folks are always watching. If the behavior you are looking for is adult and mature, then your actions need to elicit that type of behavior. Dictating to people really gets the goat of your truly motivated problem-solvers because it restricts them from using their skills to solve organizational issues. The bottom line is that, if you are a manager and want your folks to view the world as "the commons," you must give them room to problem solve,

> The additional aspect of confidence that you should consider is that your organization doesn't care if you lack confidence. They are looking for leadership and, in that regard, expect you to take on tough problems.

provide input, make decisions, and be accountable for the results. You, in effect, must be the ring master and know when your part is under the spotlight.

Taking Credit

News flash, chief officers: it isn't about you. That said, often the heads of organizations take an upfront approach to be the face of the organization. I am sure you can reflect on sports franchises, large corporations, and even small businesses where the name of the organization and the name of the CEO are synonymous. In many of these situations, publicity and the charismatic nature of the leader are what sells the product and therefore creates additional business.

The difference between us and that example is the difference between service and sales. While one could advocate that we do sell a product in the fire and EMS business, that product is performed as a service to our citizens, and additionally most of them have no choice but to accept what product we provide. In most examples where a product is being sold, the purchase of the product is discretionary on the part of the buyer. The point to all this is that branding and recognition are very important to those businesses that provide products when the customer has a choice in the purchase.

As was stated earlier, being the confident leader means that you will be out front and guiding your organization toward success. This frequently means that the press will want to talk with you about organizational business and fire and EMS operations. Sometimes this process plays out in such a way as to give citizens the perception that the chief is solely responsible for the results. This type of situation will have a positive impact on how your organization is viewed by the public but may have a detrimental impact on the very deserving employees that caused the success to begin with.

In most instances, the success of your organization can and should be attributed to your employees, to the extent that your involvement in the success should be minimized. The problem for many of us leaders is that we have egos, and sometimes those egos can cloud our judgment and our willingness to give credit to others. Let us be honest, most of us worked hard to get to this point in our career, and a little recognition always helps boost our morale and confidence. In this regard,

> The problem for many of us leaders is that we have egos, and sometimes those egos can cloud our judgment and our willingness to give credit to others.

we put our pants on the same way as other employees and have the same personal frailties.

The point here is that unless you are wearing your turnout gear and a breathing apparatus, running a 12-lead defibrillator, or working an extrication tool on an entrapment, you are not the one doing the work, so how do you possibly decide it is important to take the credit? A good rule of thumb for a chief is that if you did not do something by yourself then do not take credit, and even if you did do something by yourself, you should reflect on the fact that it may not have been possible without someone's previous effort. That previous effort could have been a training opportunity you were afforded or the personal mentoring and coaching given to you. There is almost always an opportunity to give credit to others for our personal and organizational successes.

Giving credit to others is one of the most powerful leadership tools that chiefs have in their bag of leadership strategies and tactics. You see, your job as the leader is to build up those around you. You should view your job each day as just another opportunity to help others grow in a way that will benefit them both personally and professionally. At the end of the day, it is about placing other people on the platform of success that will determine how you are remembered. How many lives do you touch by your caring and compassionate nature? How many officers did you create for our fire service profession? How many people did you give credit to when credit needed to be given? These are all important questions that leaders ask themselves prior to speaking at a funeral. You see, the answers, and the positive results that come from them, are much more powerful when the recipient is alive.

As an operations battalion chief, I oversaw a routine room-and-contents fire that was venting through a front window. When we arrived, the press was already on the scene, so I remember thinking about all the things that could go wrong and would be documented forever in video footage. You know what I mean: those calls where the driver falls out of the truck or the supply engine does not secure the supply line and drives off, dragging your 5 inch hose down the street. Even worse is when you have one of those fires where the attack crew is not very aggressive and each time you inquire about progress, they tell you they are getting on it. The only problem is that you are looking at a size 10 boot in the front doorway and a hoseline that has not moved even an inch the last few minutes. Meanwhile, your room-and-contents fire is now down the hallway and up in the attic, and the remainder of the job looks like a *Three Stooges* episode.

Fortunately, that did not happen to us that day as the operation went about as smooth as it could go. About the time I got the command board up and started accountability, the steam from the attack line was venting through the front window. The incident was a great stop, and even better was that it was all captured on video.

The operations district chief stopped by the fire as well, and as soon as the media saw him, they went immediately for an interview. Not being rude, he said that he would check with Command and then be back in touch in a few minutes. After a briefing from me and my positive comments concerning the operation, he inquired as to which company made the attack. He then gave a very professional frontline suppression captain and his crew the first opportunity to be briefed by the media.

I recall how proud that captain was that his family would see him on television. Can you imagine how powerful a leadership moment it was that the credit for a great outcome was given to him and not some chief officer with a white shirt and black tie? No, instead credit was given to the person who crawled down that hot and black-smoke-filled hallway and did the job.

> The reverse of accepting credit when things go well is the leader's willingness to accept responsibility when things don't go well.

The lesson in this for me was very powerful. First, as the leader, the district chief gave credit where it was due, and that was to the folks who performed the work. I am positive they considered the many hours of training leading up to that event all worthwhile. Secondly, the message to the organization was that chiefs were the ones least responsible for the successful outcomes on the fireground. From that point forward, I have always given credit away and shunned accepting it when it was forced my way.

The reverse of accepting credit when things go well is the leader's willingness to accept responsibility when things don't go well. Failure and periods of disappointment are the very times when the leader should step up and be the face of the organization. Your job in these situations is not to deflect the criticism or make excuses if it is justified. Rather, your job is to accept the criticism on the behalf of your folks. Under no circumstances should you throw one of your folks under the bus or have them take one for the team. Your job as chief is to take one for the team.

I have witnessed great leadership in my time in this profession, and I can say without a doubt that the best and most effective leaders always deflected the light of success when it's shined in their direction. The ironic part of this type of behavior is that being humble and giving credit to others provided even more opportunities for recognition. Those increased opportunities for recognition in turn offered those leaders even more opportunities to give credit to others. There is one powerful lesson in this type of leadership behavior.

Authority and Control

At work, just like home, I am not really in charge of anything. People may call me the chief, but I am more responsible for things than in charge of them. Heck, the truth be known, I am probably the least responsible for most activities that occur daily in my organization. Sure, I may toss a "we need to go this way or that" out there, but regarding the bulk of the service delivery, I am very far removed.

Are you a control freak? Do you view yourself as the most important and powerful person in your organization? Do you view yourself as part of the team or over the team? Do you need to know everything that is going on or just those things that apply to your organizational responsibility? Perhaps a better question is how do your people view you when it comes to authority and control?

When it comes to leading an organization, it is best to view yourself as the steward and not the dictator. The steward in this case just guides the organization toward success and is the temporary caretaker of that responsibility. The dictator or authoritarian, on the other hand, must direct things and be in charge. The whole dictator thing has not really worked out well for those folks who have tried it before, and that can be said for leading small organizations or large nations.

During my time in the fire service, I have seen lots of change. That change includes shifts in culture that make up the individual generations of our workforce. For example, when I started in the fire service, the old-timers of the day came up in a system where you did nothing until you were told, and after that you waited to be told to do the next thing. My generation had a few more questions, but at the end of the day we also did what we were told. The current generation of young men and women taking the reins of leadership in the fire service are questioning everything, and I think that is very refreshing. I believe this recent change in culture will be most beneficial in helping the current and future fire service generations deal with the rapid pace of change.

> Leadership that displays "results orientation" stays focused and is committed to doing everything possible to keep the organization directed at achieving the mission.

At one time in our history, being a one-man band fire chief was perfectly acceptable with employees, as they were very comfortable waiting to be told what to do. The pace of change was also much slower, and therefore leadership had time to evaluate the potential pros and cons of changing methods or procedures. That is not so much the case these days.

Successful organizations these days empower their frontline employees to make "on the fly" decisions that affect the way services are delivered. This is essential for several reasons, the least of which is that the pace of change today is faster than any one person can keep up with. One person calling the shots, or even thinking they can call all the shots, is just not an effective or modern leadership practice.

> Charisma can also garner you support with employees when it is demonstrated by participating in organizational activities as a part of the team and not just the leader of the team.

The relationship of the leader to the worker has also changed in recent times. Past leaders were able to direct employees by telling them what the right thing to do was, while the effective leader of today is asking employees what they think before making decisions about organizational direction. This "what do you think" kind of feedback is not perceived by current generations as a weakness on the part of the leader; rather, it is seen as a strength.

In today's fire service, it takes a team of people all committed to achieving the organization's mission for change to be managed effectively. That means many levels of decision-makers whose opinions about how things should be done are not only respected but trusted. If you fancy yourself as a dictator or the big man in charge, it will only be a matter of time before your employees will be standing around waiting to be told what to do. At the end of the day, you will find that this is a very dysfunctional way for an organization to run. We will talk more about methods and practices that create organizational empowerment later in the book.

Success for the effective leader is less about your control and authority than it is about the control and authority you give to others. Since even the best of us cannot push or pull the organization by ourselves, it is imperative that we build and maintain relationships so that all members have an appropriate level of control, authority, and responsibility. When this kind of environment is created, you will be amazed at how well your employees perform and the great things they accomplish.

> Successful organizations these days empower their frontline employees to make "on the fly" decisions that affect the method and way services are delivered.

The effect of excessive control in organizations is that it creates boundaries. Think about your own organization's policies and procedures for a minute. Defining what is acceptable limits employee behavior to those things that fall inside of the narrow boundaries. When this control is excessive, and employees feel trapped inside the boundaries, the result is resistance. I am not advocating those employees be left to their own devices here, only encouraging that as the leader you need to understand how excessive policies create control that ultimately stifles employee initiative.

As the leader, you will be given authority to make decisions concerning the organization and employees. In this regard, you will be required to discipline employees for misguided behavior and correct employees when they are headed toward mistakes. Just remember the hammer you have is a big one and can inflict tremendous damage.

> Organizational humor is a different animal. These situations do not involve the camaraderie of a team in battle but rather can involve very fragile feelings of coworkers, many of whom don't understand a life-and-death situation of a high-stress environment. These are the situations that get our folks in trouble because the battlefield does not translate well to the boardroom.

The way you approach these discipline opportunities is critical. Each one of them can be viewed on a scale from "How dare they violate my policy" to "I'm disappointed because they let the team down." In the first instance, the message is that you are in control of them and they are less than you, while in the second, the message suggests that you are all on the same team. You have a responsibility to discipline, but when doing so you are no more important to the team than they are. It is the same hammer, just a relationship-based approach to how you swing the thing.

Finally, please keep in mind that your authority as a leader is given out of respect and it is something you can lose very quickly if mismanaged. Being an authoritarian, control freak, or otherwise "my way or the highway" kind of leader is the fastest way to lose employee respect.

Delegation

When I first started in the rescue business, I was mentored and coached in rope rescue by the late Mike Brown. As I previously mentioned, Mike was the rope guru of Spec Rescue and went on to publish a book on the subject. I mention Mike here in this section not so much because of his expertise in rope rescue but rather because he was a great instructor and mentor when teaching others rope rescue techniques.

Mike was a very commanding presence. He was a big dude with a deep and very loud voice. When Mike spoke, he got your attention and then because of the expertise held you captive to what it was he had to say. I loved teaching with Mike because he presented himself as a professional instructor and expert technician and in doing so pulled all of us sideliner people on his coattails.

The peculiar way that Mike taught other instructors was to make sure they understood the techniques and concepts, and then he would place them in situations to stretch that capacity ever so slightly. I have come to realize that one of Mike's best attributes was understanding that when people are forced to teach something, they are placed in a situation to learn it again. They not only teach the students but reinforce the skill in their own minds at the same time.

In Virginia Beach, we did most of our rope training on the Seal Team Tower at Little Creek Navy Base. The tower was only four stories, tall enough so most people could learn rope skills without the fear of dying. We almost never mentioned that four stories was just tall enough to ensure that if something did go wrong, they would most likely be paralyzed. For the record, we never paralyzed or killed anyone. Anyway, the tower had a huge platform and wall structure that would comfortably provide room for 40 or 50 students.

One day on the top of the tower, we were teaching mechanical advantage skills, and I was assisting one of the other instructors at his station. Not having all the confidence in the world, I normally listened very carefully, watched the students perform the skills, and then, where needed, helped them through various difficulties. Mostly I just looked for stuff I knew for sure I could help with and ignored the skills I had less confidence I could teach (fig. 6–3).

FIGURE 6–3. Author doing Virginia Heavy & Tactical Rescue Team Special Operations Rope Training

On this day Mike seemed to be watching me more closely than usual. Finally, he said to me, "What the hell are you being so quiet for?" My response was that I did not realize I was being quiet, as it just didn't seem like there was anything to interject in the conversation. To this he looked at my station lead instructor and asked him to go help at another station that was falling behind. At the time, my immediate thought was, *This is stupid, won't our station fall behind now?*

As my lead and I starting walking toward the other station, Mike looked over and said, "Martinette, let Dave go and help out the other station. You take over here." *Son of a bitch*, I remember saying to myself. I was now the station lead. Now there would be no more comfortable place to just stand and listen.

Obviously, everything turned out fine, and I was able to teach the students the necessary skills. This story shows how Mike always found a way to push me to an uncomfortable place all the while knowing full well that I could perform. He would provide stretch opportunities for me while making sure that I had the skills to accomplish them and not fail. I will always appreciate that about Mike.

When you delegate, do not just delegate to give other people tasks that you don't want to do. You should only delegate when it is organizationally necessary because you have more strategic or complex issues to address or the tasks associated with the delegation are not your role to begin with. Remember that as the leader your job is to get things done, not necessarily to do all things.

There is also a way to delegate to help people grow, the same way as Mike did. Mike used delegation to push people's comfort zones by assigning them tasks that they could do but that also stretched them to the point they learned new things and were challenged in the process. This is how high performing organizations succession plan for their organization's future.

As you move through your leadership career, it will be very important for you to develop good delegation skills. This is because delegation in high performing organizations is a tool used to encourage and nurture employees to be better while at the same time moving the organization forward in the most efficient and effective manner. In the process, you assign tasks and projects to people based on their strengths and abilities to ensure that they will be successful and at the same time grow as an individual.

Another aspect of delegation consistent with high performance organizations is delegating tasks that you like to do but do not really need to be doing. All of us have the jobs we like to do because they are easy and fun. The problem is that most of those are jobs that should be done by other folks in the organization. Do not get caught up in a vicious cycle of working on all the fun things at the expense of the truly important work your organization needs you to do. If you find yourself telling everyone that you just cannot seem to find time for this or that or you can't seem to get out of the office on time, I will bet my last dollar you are working on things that are fun, easy, and uncomplicated that, when all is said, other people really should be doing.

> When you delegate, do not just delegate to give other people things that you don't want to do. You should only delegate when it is organizationally necessary because you have more strategic or complex issues to address or the tasks associated with the delegation are not your role to begin with.

Mike Brown was a great instructor and a true pioneer in our business. As he recently passed away, we now have a huge void in specialized rescue training. Rest easy, Big Daddy, I miss you.

Happiness

If there is one trait that gets me in the most trouble with firefighters, it is happiness. I am not generally an unhappy person, but it is the mentality of some firefighters that my sole purpose in life should be to assure their personal happiness.

If you are a fire chief, you are laughing right now. Conversely, if you are a firefighter, you may think my statement is somewhat demeaning. Nothing applies to everyone and every situation, but it can on occasion appear to leadership that they are stuck on a hamster wheel of trying to satisfy whatever is perceived to be the next thing wrong with the organization. Believe me when I say there will always be a next thing wrong with the organization.

In the fire business, we are quick to bring up the term *morale*, the perceived emotional well-being of the workforce. Most commonly the statement is conveyed as "Morale of the workforce is terrible" or "We have terrible morale in the department." The funny thing about this word *morale* is that in my 45-plus years in the fire service, I have never heard the term used with anything other than a negative connotation. For instance, filed under the subject line Never Spoken Around a Firehouse Table is "The morale of the department is great" or "I have never seen the morale of the workforce any better." It just does not happen in the fire service because, as we previously mentioned, the nature of firefighters is to concentrate on "broke stuff." However, if you stay focused on "broke stuff," everything looks broken.

> The funny thing about this word *morale* is that in my 45-plus years in the fire service, I have never heard the term used with anything other than a negative connotation.

So, what is a leader to do about this situation? If the nature of our employees is to always look to fix something and therefore always perceive the workplace as broken, then what is our chance of being successful as leaders in trying to create an enthusiastic and motivated workplace? In my mind, it starts with how we talk about happiness and morale.

I can vividly remember being a young man of 9 or 10 and being upset about my mom's behavior, the way she would manipulate a situation to make you feel one way or another depending on her personal motivation at the time. For instance, if she wanted me to be mad at my dad, she would create a situation in which it would look like he would disappoint me, like telling me he was coming to pick me up when that was never the plan. As you can imagine, a young child can feel very disappointed when he thinks his dad is coming to get him but then his dad does not show up.

It turns out that people with her illness may display this type of action as one of many behaviors that are detrimental to building strong and bonding relationships with others. As I mentioned earlier, I do not under any circumstances harbor ill feelings about the situation, as I truly have come to understand that people with

> Regardless of how desperate things might be, or may appear to your employees, as the leader you absolutely must hold out hope that things will get better. If you want to watch a slow-motion leadership train wreck, just watch a leader during a crisis who loses hope.

mental illness cannot help themselves any more than folks with cancer can cure themselves.

No other human being on earth is responsible for your happiness but you. Sure, someone can create a situation where you have the potential to be unhappy, but the choice to be happy or unhappy is very much your own. Nobody else, and I mean nobody, can control your choice to be happy.

Make no mistake about it: some people are born to aggravate the crap out of you. It may at times appear that their sole mission in life is to create situations where you let yourself be unhappy. The point is they are not creating your unhappiness. They are just creating the environment where you are letting yourself be unhappy.

This is not to say that you will never be unhappy. We all have events in our lives that cause us to be unhappy. To be unhappy at times is to be human, which is a characteristic all who are reading this book share.

As a leader, your employees want you to be happy and to create an environment where they can be happy. Nobody wants to work for a pain-in-the-ass boss who is in a bad mood all the time. If you are unhappy all the time, what can you expect from your employees but the same behavior?

Remember these points about happiness: First, you are the caretaker of your own attitude. Second, if you are not happy with a situation, then remove yourself to a situation where you are happy. Third, if boredom is causing your unhappy nature, then find something else to do that energizes you and makes you feel good about yourself.

Courage

Make no mistake about it, courage is an attribute that defines success in the firefighting profession. From making your way down a hot hallway, to making fireground decisions where people's lives are at stake, to deciding on issues in the absence of defined policy, it takes courage to be in our profession.

> The idea behind these playbooks is to outline instances where you want or need consistency of action. In other words, "When this happens, do this."

Great leaders are created on the street. Those "moment of truth" times in our lives where firefighters make decisions outside of orders form a solid platform for understanding what the right thing to do is and what could ultimately be a career-ending decision.

For those of you not in the fire service, you should know that we do have a playbook of sorts. That playbook is called standard operating procedures (SOPs) or standard operating guidelines (SOGs). The idea behind these playbooks is to outline instances where you want or need consistency in action. In other words, "When this happens, do this."

For reasons of life safety, these playbooks are very important and serve a valuable function. For instance, when an engine company needs to advance a line in a house fire, they need the truck company to provide forcible entry into the building. Once inside, they may need the truck personnel to ventilate the structure or conduct flow control so they can make it to the second floor back bedroom. In this manner, each member has a job and every other member depends on them doing that job so that everyone is successful in what they are supposed to accomplish.

To be upfront here, it does not take a tremendous amount of leadership to follow scripted directions. It does, however, take tremendous leadership to make decisions outside of those scripts. Most of the time these are the types of decisions that also require tremendous courage.

As a recruit firefighter, my engine company was assigned a box for a fire alarm at a local strip mall. This place was a new restaurant that had significant renovations completed in the last few weeks, and because of that we had run several alarms of the same type. These automatic alarms are an important part of protecting the public and keeping people safe; however, that is a hard sell when it is 4 o'clock in the morning and you have been on five of them since midnight.

Back then these types of alarms were only single-unit responses, and so if something was going on you needed to call for additional units. On this occasion, we arrived and gave our initial size up as nothing showing. Nothing seemed out of the ordinary.

> The success of most organizations is almost exclusively judged on how they manage the pace of change and, maybe even more importantly, how they organize and operate to get out in front of it.

Upon exiting the truck and starting our walk around, I happened to notice a slight smell of gasoline or some other similar petroleum product. Upon further examination, the front windows were darkened and appeared to be glazed with spider-like cracks running down their length. We also noticed an ever-so-slight wisp of smoke coming out from around the rear door. I remember thinking, *Holy crap, I think we have a fire here after all.*

At that point, I was all for calling in the cavalry, so to speak, but my officer wanted to do a little more investigation into the situation. In our business there is nothing more embarrassing than calling in a job and then it turning out there's nothing to it. By the way, this is a situation that happens to all new officers at least once in their career.

"Pull me a line, Rookie," my officer told me, and with this direction I grabbed the preconnect line and pulled it toward the front door. When I was about 10 or 15 ft. from the front door, my officer took his Halligan tool and speared the area between the door jam and the door itself. He pulled back as hard as he could, and the door popped open, and much to our surprise the place was burned out. It looked like someone burned the crap out of some furniture, then placed it in the restaurant, closed the place up, and then called us just to see what our faces would look like when we opened the door. I can tell you one thing: mine was surprised.

> As it turns out self-discipline is a little like accountability. Everyone can agree that it is easier to expect these kinds of things from others than it is to expect them of ourselves.

What made this situation unusual was that the smell of gasoline was unbearable. It was so thick and noxious that it took both of our breaths away as we fell to the ground and put on our breathing apparatus. My officer said, "Follow me, we've got to get this place ventilated and get it done in a hurry." To this end, we both crawled from the front door to the rear and opened all the doors along the way. When we retreated out of the back door, both of us were soaked with gasoline and would have ended up like two roman candles had there been a spark in the area.

On many occasions, I have wondered how close we were to becoming enveloped in fire at that restaurant. The timing of forcing that door must have been perfect, and the fast ventilation of the building must have been just enough to keep things from lighting off. As luck would have it, the fire had flashed and consumed all the oxygen in the building, which limited its involvement. In addition, there was too much gasoline in the building, making the environment too rich to burn. In the end, everything turned out just fine.

Now I will be the first one to tell you that in most cases it does not take much courage to follow directions. Orders from your officer do not normally come with a "How are you feeling about this one?" conversation. But from the other perspective, someone must make those orders, and the consequences can mean life or death. In our case, my officer decided to force that door and aggressively get the building ventilated, and in doing so we preserved valuable criminal evidence that was used to convict the person that tried to burn the place down.

Just imagine for a minute what kind of scrutiny my officer would have been subjected to if the place lit off. You can bet there would have been much criticism of our actions leading up to the event. Critics would have argued that his decisions were not appropriate and that we should have waited of other units to arrive before initiating actions. It took a tremendous amount of courage to make the decisions that ultimately led to our success that early morning: personal courage, in that his life was at risk, and professional courage, in that his career could have forever been defined by this event.

My officer did not have the luxury of hindsight to help make a series of decisions to stabilize that incident. No leader ever has the benefit of hindsight, and that is where leadership courage is so valuable. Effective leaders have the courage to make decisions and not be paralyzed by fear no matter whether that fear is for their own safety or for the safety of their jobs.

> No leader ever has the benefit of hindsight, and that is where leadership courage is so valuable. Effective leaders have the courage to make decisions and not be paralyzed by fear no matter whether that fear is for their own safety or for the safety of their jobs.

During the early days of the Murrah Building bombing incident, there was reason to believe that what was left of the building was going to further collapse. While working in the basement to recover remains, our videographer noticed a separation in one of the large beams that was supporting a majority of what was left of the building's basement. After a quick review of the video he shot the day before, it was concluded that his observations were correct. The beam was indeed separating from the columns that supported several floors in that part of the structure.

It was decided that we would retreat to the safety of the base of operations until an analysis of the conditions was provided by our structural engineers. The engineers were then expected to develop a plan regarding the safety of the building. Keep in mind that engineers are more familiar with determining the safety of a structure when all its components are operating as intended and not when they have been blown to pieces. I remember all of us thinking to ourselves, *Of course the building is not safe. A bomb went off here, remember?*

One of the Incident Support Team members who oversaw basement operations was FDNY Rescue 3 Firefighter John O'Connell. I knew John through his work in developing the shoring modules for the FEMA Urban Search & Rescue in the Rescue Specialist certification program. John, along with a very select group of firefighters from around the country, was the expert on shoring operations for structural collapse incidents.

During a protracted meeting to discuss our next steps in the rescue and recovery, it became very clear that no one person, including the incident commanders, was willing to step up and decide. That was, until John spoke up.

I think it would be safe to say I had more confidence in John than I had in our engineers. I mean, John had actual experience working in buildings that were compromised, and these other guys just designed and built buildings. His opinion was that we were wasting valuable time and indeed the building could be shored up with the proper materials and expertise. That, however, would not be the case much longer should we all decide to just stand around and do nothing.

As it turned out, John was right. About 25 of us entered the building that night, and with the winds blowing and the building swaying, we proceeded to shore up the basement. When we were done that next morning, the area was nicknamed the "forest" because it had so much wood down there. When the building was imploded many days later, the demolition experts said it took more explosives to demolish the "forest" than for the remainder of the building.

I took several lessons out of that experience. First, John was willing to put himself on the line. He had the courage to step up even when this decision was questioned by others who had more formal education in structural engineering. Note to self: education is a very valuable tool but can in no way be a substitute for real-life experience. The second lesson for me was that the members of our team followed John into that building based on his personal commitment and courage. He did not just convince people what needed to be done but also led the effort. He had just as much

> Being an effective leader is often realizing that success during and after some adversity rests in how we act while it is occurring and how we learn the lessons it is teaching us.

on the line as those of us who followed him into the building that night. Virginia Beach Task Force II got most of the credit for the "forest" and what it did to stabilize the building so that all remains could be recovered; however, John was the one responsible for us being there in the first place.

As a fire officer's career develops, literally thousands of these operational lessons learned get filed away to be used in instances where operational problems are the farthest thing from the actual situation. Rather than managing fire situations or shoring operations, fire officers deal with politics and doing the right thing in situations where personal courage has more to do with personal job security than bodily safety.

As a chief, I have many years of budgeting experience. During these times, a chief can get caught between what their manager would prefer in a lean request and what their folks want regarding equipment and resources. This push-and-pull scenario plays out across this country and frequently leads to disagreements between city leaders and fire chiefs.

In 2011 the *New York Times* decided to do an article on my city that was supposed to be an account of how a medium-sized urban/suburban city was dealing with the extended years of the recession. When the *Times* reporter spoke with me, I indicated that the city had been very responsive to the needs of the fire service but that for the first time in a long time there was open and ample discussion regarding the reduction in workforce. This was taking place because of 3 previous years of reductions in operating cost. Like most fire departments, operating cost only represented between 10% and 15% of the total budget, with personnel cost in fire departments always making up the bulk of the budget.

> Great organizations view missteps and disappointments as learning and teaching opportunities. Remember that each one of your employees is watching how you as the leader behave during times that are bad and coincidently when real leadership is required.

Since the operating cost had been marginalized the last several years, there was discussion of personnel cost and potential impacts of a reduction in workforce. This was the first time I could recall that the environment was safe for elected officials to even mention a reduction in firefighting personnel, as even citizens wanted to know what the consequences of a reduction in force would mean to them and their insurance rates.

During negotiations for the previous three budgets, in addition to a reduction in operating expenses, we had a piece of equipment that had been pushed out of the Capital Improvement Plan. This is the plan used to purchase large pieces of equipment and build municipal infrastructure. In a nutshell, the city generally borrows the money by issuing debt and then pays the debt down over the anticipated life of the building or piece of equipment.

This piece of apparatus was far beyond its normal operating life and was costing us thousands of dollars a year to just keep it in service. As a matter of fact, I determined that the potential debt payment to our fleet fund on the truck was going to be very near the anticipated maintenance cost. In my mind, that alone was reason enough that the truck should be approved for replacement.

The last straw for me was when the truck stopped working during a fire operation, which placed my employees' lives in danger. I know firefighters can at times be unreasonable with their expectations of the department's leadership; however, leadership providing them with the tools and equipment to do their job safely should not be counted as an unreasonable demand. I made up my mind at this point that something needed to be done.

When the draft budget was printed, sure enough my truck was pushed out another year, and to me that just was not acceptable. Something needed to be done, and it was up to me to figure out how to get this item back into the budget discussion.

I decided to talk with one of my council members, whom I considered to be very analytical. As a retired lawyer, he was used to dealing with data and examining even the smallest of details for relevance to the situation. I knew that if I could get the discussion concerning the truck to come up during the council budget work session that the facts alone would drive the successful inclusion of the truck in the budget.

I did consider how this tactic might affect my career, as there was certainly the possibility that I could lose my job. I work at the discretion of one person, who may be offended by my actions. But I also work for and have a responsibility to each man and woman who makes up my department, and if something were to happen to any of them because I didn't have the courage to advocate on their behalf, I would never forgive myself.

My truck did come up at the next council meeting, with four of seven council members voting to approve it for inclusion in the budget. I was happier than my boss when the meeting ended. The situation created an uncomfortable next couple of months as he considered my efforts to have violated a trust between the two of us.

> Great leaders understand their role in the decision-making process and have the courage to take the next step, open the next door, and continue the operation in the face of fear and uncertainty, and in doing so they do not use personal or professional safety as the only filter in making the decision.

Over the next few weeks, we talked things out, and in the end, I think he fully understood my position, although I can say with a great degree of certainty that our relationship from that point forward was never the same. I did truthfully tell him in no uncertain terms that when I go to sleep each night, I worry whether I have provided my folks the tools and equipment to do their jobs effectively and safely. I take that responsibility very seriously.

The *New York Times* article came out, and much to my dismay it was not as much a story about our city's budget troubles as it was a story about a fire chief's plan to get his fire truck purchased. It was a shame it was viewed that way, as our city had done a remarkable job over the recession years to cut expenses without a substantial nonvoluntary reduction in the workforce. A fair number of very talented and generous managers did not get their due because of my fire truck. I was always sorry about that part of the situation.

When I took my first chief's job, I called my former Chief Harry Diezel asking him for that one piece of advice that would help me create a successful career as a fire chief. His advice to me was to never take a job I could not afford to lose. When I asked what he meant by this statement, he said that there would come a time when someone would ask me to violate my values as a condition of employment. He also told me that one day I would need to advocate for my folks to the degree that it may in fact jeopardize my career (fig. 6–4).

> Positive patterning of your behaviors by someone else is the ultimate form of gratitude.

FIGURE 6–4. Fire Chief Harry Diezel taught the author about responsibility and advocating for his members (courtesy of Ray Smith).

To his credit, both have occurred and because of his advice the outcome has been positive. I am not sure that would have been the case if I feared only for my own safety during the process.

It takes a measure of courage to act when the outcome of the situation is unknown. Great leaders understand their role in the decision-making process and have the courage to take the next step, open the next door, and continue the operation in the face of fear and uncertainty, and in doing so they do not use personal or professional safety as the only filter in making the decision.

The folks who work for you expect you will have the courage to reasonably advocate on their behalf. To do less is unacceptable.

Followership

I most appropriately and very purposely left followership as the final trait. If there is any one thing I would like you to resonate with, it is the notion of being a follower.

Many of us believe that when we become leaders, our job is less about us following others and more about others following us. To this I say that we all work for someone, and if you are unwilling to follow that person, assuming that person is a good human, then you are not a good leader.

As you go about your daily chores of leading, I would suggest you pay particular attention to how you respond when placed in the followership role. All leadership is first about being able to follow. End of story.

Exercise

Take an opportunity to do the exercise in Appendix B. Getting feedback on your traits from relatives, friends, and coworkers can help you broaden your perspective on personal leadership.

Values and Traits: Pulling It All Together

Just like individual components in a good soup, the correct proportion of values and traits is a recipe for good leadership.

Key Points

- Leadership is many things.
- Successful leadership is dependent on the proper application of many values and personal traits.

So far, we have talked a great deal about personal leadership. There are many values and traits that, when applied in the right proportions and at the right time, makes a successful mixture for great leadership.

The world around us holds some remarkable lessons in leadership if we just pay attention—if we open our minds and contemplate how interactions occur and why events unfold like they do—and then look for consistencies that contribute to the success or failure of these leaders. For good reason, smart leaders seek to avoid pitfalls and examine potential weaknesses in themselves by comparing their behavior and interaction with those who have demonstrated success.

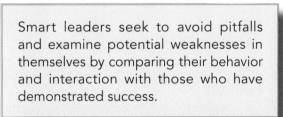

Smart leaders seek to avoid pitfalls and examine potential weaknesses in themselves by comparing their behavior and interaction with those who have demonstrated success.

Lynchburg is a small community of 65,000. Because of its size, it does not take long for the chief to learn who the influential people are. Some are influential because of their wealth, others because of what they do for the community. This story is about a man who is influential for the latter reason.

Tom Gerdy was a local Lynchburg building contractor who was involved as a volunteer for our local chapter of Habitat for Humanity International. Habitat for Humanity is a nonprofit, nondenominational Christian housing organization that organizes volunteers to build simple, decent, affordable houses in partnership with those in need of adequate shelter. Since 1976, Habitat has built more than 125,000 houses in more than 80 countries, including some 45,000 houses across the United States.

The fire department partnered with Tom, a local church called Victory Christian Fellowship, and the police department to build Victory House, a house that would be used as a haven for citizens and visitors who fall victim to an unfortunate accident, fire, or crime. This idea was an outgrowth of our attempt to provide customer service in the form of restoration and was a part of our overall fire department strategic plan, in effect helping our customers by restoring their sense of well-being.

Knowing When to Follow

I first met Tom in a meeting with the other key stakeholders discussing the overall vision for the Victory House project. He listened very intently to our conversation to see where he could fit in.

I would characterize Tom as quiet, unassuming, and slightly withdrawn. In my estimation, he was looking for a leadership opportunity; however, he needed a clear vision of what the opportunity was and where his skills would bring value.

The conversation continued as Tom carefully took notes and sized things up. At this meeting, he appeared to be anything but a leader. He was looking to us for guidance and direction and he appeared very comfortable in this role. Our meeting ended cordially, with some specific objectives documented and clearly aligned objectives determined.

Lesson

Good leaders listen before they speak so that they can most appropriately determine how they can add value. All our roles are first followership in nature, and not until after obtaining a clear understanding of the vision and objectives does the leader begin to emerge. The bottom line is that the good leader leads when it is time to lead but is a good follower when it is time to follow.

Taking Charge

The next time I saw Tom, we were at the construction site. Much work had been done up to this point, and Tom had a lot to do with that accomplishment. He had organized plans, recruited people, and done the legwork in preparation for that early morning when 50 people of varying skill levels gathered to build a house in one day. Tom caught the vision and then went to work. Once he had the objectives clearly outlined, he did not need anyone to tell him what to do or when to do it. He just did it, and we all recognize this as a basic trait of a good leader.

Lesson

Good leaders can sense when it is time to lead. They assess the situation based on their expertise and the willingness of the people with whom they are working. Once clear objectives have been established, the leader is self-motivated to gather resources and make the vision a reality.

Credibility and Trust

Tom has abundant credibility when it comes to building houses. He has built more than 70 homes for Habitat, and in addition, he maintains a successful construction business of his own. He is a fair man with an incredible work ethic and keen sense of community spirit. Tom does things for the right reasons and people understand this very quickly.

Lesson

The ability to lead people is based on their perception of your credibility and how much trust they have in you. In my experience, leaders do very little leading until they establish a level of credibility with employees and the employees trust that they are being led in the right direction. For example, the team may ask, "Why should I follow this person when I am not certain they are capable of being successful?"

> Many successful people lose their ability to lead because of poor ethical judgment. Just one lapse in ethics on the part of the leader may result in loss of the credibility needed to effectively lead the organization.

Integrity is the corner stone of credibility. Employees recognize very quickly those people who gain credibility and achieve goals for the wrong reason or by unethical means. Many successful people lose their ability to lead because of poor ethical judgment. Just one lapse in ethics on the part of the leader may result in loss of the credibility needed to effectively lead the organization.

Preparing to Lead

When we arrived in the morning to start construction, it occurred to me how much preparation had been done. Jacks had been built and numbered, as had the smaller wall sections, such as closets. Materials were arranged on the site based on when they would be needed. The equipment to build was on-site, and so were the people. All the potential barriers to failure had been assessed and addressed.

Lesson

Leading is hard work. It takes hours of preparation to effect change in any organization. Recognizing obstacles to success, gaining the support of key people, and finding necessary resources are all done in preparation for leading. In our department, we call this Completed Staff Work.

Tom had done this before; his experience helped him prepare. Therefore, when it was time to lead, Tom had completed the preparation phase, and all that was left was the work of bringing about change.

Recruiting Key People

As the work started, it became obvious that we had several experienced people among our group. Some were from the fire and police department, but others were people whom Tom had worked with before. These were folks who knew how to build a house, run a saw, and hammer a nail. The key people were experienced in building houses and knew the answers to the "What comes next?" questions the inexperienced folks asked.

These players were recruited because of their experience, and it was clear that Tom effectively used them to develop momentum for the project. Making sure they were divided based on job task then became a matter of organization. While participating in the work, Tom became the overall big picture organizer for the project and allowed the key folks to drive smaller objectives to keep things moving.

Lesson

Even the best leaders will realize only limited success if they do not have quality personnel. An effective leader needs to concentrate on working in the strategic portion of the time allocation model. Had Tom gotten bogged down in an operational detail, the overall project would have suffered.

> An effective leader needs to concentrate on working in the strategic portion of the time allocation model.

The leader needs to have key motivated people who can carry the operational work. Identifying these folks and getting them excited about your vision then creates momentum. If enough momentum is created, then the proverbial ball starts rolling, and generally only a significant negative event can stop progress.

Right People in the Right Place at the Right Time

Tom was the master at making sure that not too many inexperienced volunteers were assigned to any given part of the project. To spread out the wealth of talent, he made sure that folks who could hammer had a hammer in their hand. There is no better illustration of this than when the roof trusses were raised and put in place. Although very little skill was required to lift and place the trusses, experienced people needed to be on the roof when it was time to nail them in place and make sure they were squared up.

Lesson

Not all people in the organization have the same skills. The effective leader makes sure that key people are in key positions at the time they are needed. A leader can grow key people through long-range succession planning; another aspect of the process is making the best out of the folks you currently have.

Effective leadership also values everyone in the organization. Even though all of us have different skill and ability levels, the leader needs to find a place where everyone can add value. In addition, the leader needs to understand and recognize all stages of work in the organization. The moral is that the guy who does rehab at a fire or changes an air bottle is important when you need a drink of water or a fresh tank of air. Great leaders praise good work at all levels of their organization.

Participation in the Team's Work

As Tom circulated through the construction site, he moved from task to task organizing the strategic portion of the work. While doing this, he also participated in the work. Whether it was nailing a stud or cutting a piece of wood, Tom was a part of the team's work. As the team leader, Tom demonstrated that he was committed to the group's work and the success of the project. Moving very quickly from one area to the other, he provided energy and enthusiasm for the entire group. Tom was not an absentee leader.

Lesson

The leader must be committed to the organization's work, and the workers need to see that commitment. Even though we chiefs work in the strategic part of the organization, it is at the street level where the work takes place. The leader needs to support the workers and frequently provide direction and enthusiasm and demonstrate energy that keeps people enthusiastic about the work's progress

Maintaining Momentum

Building a house in one day is quite an endeavor. It requires many people, all staying focused on the goal, and tremendous energy. Tom is the master of momentum. Not being one of the most talented construction workers on the site, I looked for those opportunities that provided direct supervision or those that required none. Those tasks that required very little supervision usually are communicated as "Let's move these materials and distribute them evenly around the house." For example, by distributing siding materials around the house appropriately, the skilled workers would not have to go far to get the siding when it was needed.

I wondered why this was not done during the preparation phase so that we didn't have to waste our time moving it during the project. Clearly, someone could have already moved these materials. In fact, to be quite honest, it seemed that moving these materials was unnecessary in the grand scheme of things.

Keeping folks busy maintained focus, energy, and momentum. When workers hit a lull and it was easy to lose focus, Tom had everyone moving things around. He maintained the momentum of the project during lulls in the work by tricking us into thinking we were adding to the preparation of some event that would take place in the future.

Another important role Tom played during the project was recognizing slow work and stepping in to jumpstart tasks or give groups additional resources. The effort to put shingles on the roof was a great example. We had only two people on the roof who knew how to shingle, and the rest of the workers on the roof, while full of enthusiasm, lacked experience. Recognizing that the roof teams were not making good progress, Tom grabbed a few experienced people from another area and headed for the roof. The few minutes he spent up there and the addition of experienced people got things back on track.

Lesson

Leadership is multileveled in that it takes place at the organization's strategic, operational, and task levels. The many types of activities our organizations are involved in at any one time can illustrate this. Planning committees, work groups, process improvement groups, shifts, and battalions are just a few examples. The leader is responsible for achieving objectives and helping the organization survive the change process. Compounding the complexities of leadership at these multiple levels is the fact that many of us who lead formally are involved at all the various levels.

Momentum and focus are critical if the leader is going to guide a group toward a common goal. Think about how many times you have seen project work groups that start out with a full head of steam hit the proverbial wall and then struggle to make additional progress. Perhaps it was because all the easy stuff was done first, and the people lost direction and focus after the initial burst of energy. Maybe the excitement for the project died down a bit, and people went on to other things.

Regardless of what causes the lack of forward progress, it is the leader's responsibility to maintain momentum and energy in the effort. This may be as simple as recognizing it isn't wise to pick all the low-hanging fruit early in a project or strategic effort. Save some of the low-hanging and easy jobs so that when folks are struggling you can give them something simple that

> Compounding the complexities of leadership at these multiple levels is the fact that many of us who lead formally are involved at all the various levels.

adds value to the overall effort. Keeping people moving toward the goal at a steady pace may be as important as anything the leader can do to ensure the organization's overall success.

The effective leader must also recognize that direct involvement or the assignment of additional resources to a work effort may be necessary to ensure the success of a project or organization. Workers need to see their leaders during difficult stages of a process and know that they are committed. Being there to help, reorganize, and provide an additional resource is a positive leadership behavior.

Sharing a Laugh

Shared experiences build team integrity, and nothing makes those experiences more positive than when the leader and team members have a sense of humor. Tom used various humorous situations to create a positive and relaxed atmosphere. Even though everyone was working hard, Tom realized that without an enthusiastic and fun atmosphere, the experience would turn people off. During a furious stretch and about the time we were going to raise a large wall section, Tom yelled for everyone to stop working. I could not believe he was going to take a break when it appeared all of us were focused and really driving. As I went to unsnap my nail belt he yelled, "Break's over!" That is right; we got a 9-second break. We all had a good laugh as Tom lamented that he was getting soft: breaks usually only lasted 6 seconds!

Lesson
The people who work for us will work harder if they enjoy what they do and are relaxed. A part of every leader's responsibility is to make the work environment a positive place. Correctly timed humor can do wonders for refocusing the team's work and sending a message about the work and the environment. Leaders who are uptight and serious all the time create a very uptight and serious work environment. This type of environment is not conducive to self-motivated workers taking a risk and trying new things. We certainly must be serious on occasion, and no one is advocating an unprofessional work environment, but successful leaders need to make humor a part of their life and the lives of their employees.

When Bad Things Happen

You are not going to construct a house in one day and have everything go as planned. This was true in our building regarding a missing roof truss. Earlier in the day, we had divided the trusses into two piles in front of the house. The actual placement and securing of the trusses did not take very long; however, lifting them to the roof was another matter.

Using leverage and physics to our advantage, Tom would call for a group of about five or so folks to raise one end of the truss toward the roof line while others held the peak and back of the truss down toward the ground. As the truss was raised, at just the right time, Tom would call for the group to raise the rear of the truss, using the outside wall of the house as a lever. The process made it easier for the folks on the ground and those working to secure the trusses.

As you can imagine, it took several tries for the two groups of truss lifters to get it right; this provided plenty of humorous ammunition for Tom. However, the problem was with the last truss; we were one short! Without the last truss in place, we would not be able to get the plywood on the roof and begin to paper and shingle the roof. Almost without missing a beat, Tom said, "Well, that could be a potential problem."

We knew there was not a load of extra materials on-site and that this could be a real holdup. I mean, you do not just find prefab trusses in the right size lying around on a Saturday afternoon.

As all eyes moved to Tom, we could see that he was already developing a plan for solving the problem. With a few key people and a little time to construct a truss from scratch, the project was off and running again. As with all our little unforeseen issues that day, our leader handled it with a positive attitude and a can-do spirit.

Lesson

As chief of a department, I can tell you that not all news is going to be good. Bad things are going to happen to all organizations. We certainly want more good things than bad to happen, but no leaders can expect all positive outcomes for their organizations.

How organizations rebound from these bad events usually depends on how the leaders behave and how the employees engage based on this behavior. Employees look to their leaders during times of crisis for confidence and direction.

It is also very important for the leaders to recognize emerging problems in their organizations. Effective leaders not only solve problems as they arise but also look for potential and emerging problems continually and then do something about the problems before they become career killers.

> Employees look to their leaders during times of crisis for confidence and direction.

As Tom demonstrated in the example of the missing truss, it is important not only to recognize the problem but also to communicate the problem to all the people involved. If you want employees to be engaged in fixing things, you must include them and yourself as a part of the solution.

Taking responsibility as a leader does not have to be a bad thing. For instance, when you ask employees about traits they most admire in leaders, after honesty and integrity, you are most likely to hear humility—humility in the form of the leader saying, "I don't have all the answers" and "I do make mistakes." Humility makes the leader human, like the folks they lead, and, consequently, a part of the solution when fixing a problem. As mentioned earlier, humor can temper stressful work environments and put employees at ease.

When bad things happen, it is time for the leader to be most visible. The leader's visibility demonstrates responsibility and, more importantly, shows a commitment toward making the situation better.

One thing that the leader should never do is hang their head, even for a moment. If you lose confidence in your ability or the organization's ability to rebound from the problem, what are your employees supposed to think? How do you think they are going to act? As my grandfather once told me, "The mark of a man (leader) is not created in how he acts when things are good but more how he acts when things are not going well."

Keeping Track of Timing

Any person who has built a house knows that timing is everything. Certain parts of the building process must be done before others can start. The worst-case scenario is for tradespeople such as the plumber and electrician to have scheduled your job and not be able to work when they show up. This can mean the difference between building a house in 6 months or a year. Now imagine the precise timing required when you are trying to build a house in one day!

Tom had the timing aspect almost down to the minute. The electricians were scheduled for about the time the roof trusses were put in place. It did not make sense to have them there before the framing on the inside was complete, since they couldn't run wire and all the folks on the inside of the building would have been in the way. The same was true for the plumbing, heating and air, siding, and landscape folks. All these skilled people knew when to show up based on Tom's experience and keen sense of timing.

Lesson

Timing is critical to the success of any leader. The leader needs to have a keen sense of when the time is right to introduce change or initiate new programs. In addition, timing is critical when it comes to holding folks accountable for poor performance or praising their positive contributions.

From an organizational perspective, the leader's success is dependent on the abilities of the employees. Just as is true for house construction, successfully leading an organization requires that certain things happen before others. You cannot hold people accountable and responsible for their behavior and actions unless you have previously outlined what your expectations are.

In addition to organizational timing concerns, the leader must also be aware of political timing. Things within the organization may be in desperate need of modification; however, if the political support is not there, you will fail. It does not matter how right or necessary the issue that needs to be resolved is at that time.

An example of poor political timing is the leader advocating additional staffing when budgets are tight and other areas in government are being forced to cut back. In this scenario, the leader may appeal to the politicians that it is unsafe to operate without additional staffing. The appeal for public support might sway the politicians to provide additional personnel, even when it is against their best judgement. The result is the organization will lose somewhere else, or the leader ultimately loses by having to leave. Bottom line is that you may win the battle only to have your organization ultimately lose the war.

> You cannot hold people accountable and responsible for their behavior and actions unless you have previously outlined what your expectations are.

Taking Care of Your People

Another lesson on leadership and house building is derived from the way Tom took care of the people who worked that day. He personally saw to it that each of us had food, water, and a bathroom facility. In addition, he was constantly watching out for our safety by making sure we were wearing our gloves and eye protection.

The way Tom interacted with people showed a deep commitment to the welfare of each person. He would ask how we were doing and would also stop to praise our work. He realized that to successfully complete the project he needed workers and those workers needed to know that he cared for their personal and psychological welfare and safety.

Lesson

The success of the leader depends on the followers more than any other factor. If their first concern is for their own safety or welfare, they will lose focus of the vision and objective. Maslow's Hierarchy of Needs Theory maintains that people need their basic health and security to be taken care of before they commit to higher-level needs. This established theory has been proven time and again in connection with building organizations. The leader needs to be tuned in about how to make sure the needs of employees are constantly addressed.

It is up to the leader to maintain a check on the people's welfare and to frequently engage them in conversation regarding their feelings and well-being. A big part of what the leader does is shaking employees' hands and slapping their backs. They are the ones who get things done and, subsequently, make the leader look good. They are the most important assets of the organization; without their support, the leader will not be successful.

Taking the Credit

I once watched a local news story about Tom. The focus of the story was a national award he won from Habitat for Humanity. As he was interviewed, it was obvious he wasn't comfortable with the attention. The reporter asked about the number of homes Tom had built for Habitat. His answer was that "it wasn't important."

> The successful leader realizes that success comes because of the drive and commitment of the people who make it happen. Leadership is about helping people grow and making them better because you are there.

What was important to him was that people who otherwise would not have a home did have one. He was most thankful to all the people who helped him build houses. He deflected attention from himself and directed it to those who helped him succeed as a leader.

Lesson

If you are the leader, it is not about you—it is about the people who make up your organization. Taking credit for their work just because you happen to be the leader is an injustice to you and to your people.

The successful leader realizes that success comes because of the drive and commitment of the people who make it happen. Leadership is about helping people grow and making them better because you are there.

As you can see, Tom taught all of us many great lessons in leadership the day we built Victory House. Watching him on the news, he was still teaching me. Being humble is perhaps the greatest leadership lesson of all. True leaders do not need praise and acknowledgment because that is not why they do what they do. True leaders care about people and managing their growth. It is through your employees' growth that you are acknowledged. Tom caused all of us who were paying attention to grow that day.

Leadership is many things, and successful leadership is certainly dependent on the proper application of an array of values and personal traits. Using solid and consistent values to align your behavior and appropriately applying traits at just the moment they are needed goes a long way in determining your future success as a leader.

8

Personal Leadership and Risk

The most vocal employees are always the ones with
20/20 vision once the consequences of a decision are known.

Key Points

- Leadership in organizations is risky business.
- There is a risk-benefit situation in all leadership decisions that involve change.
- Change is in general less risky as more information becomes available.
- Effective leaders modify their decision-making based on potential consequences.
- There are favorable attributes to left-side, right-side, and midpoint leaders, however there are also consequences to not adjusting style based on the situation and risk.

Richard Branson of Virgin Airlines said, "If you are not risking something then you are standing still." In the rescue business, we teach risk-benefit analysis as a tool to ensure rescuers do not unnecessarily risk their lives. The idea is to evaluate the potential for death or injury to the rescuer and then balance that against the potential for a successful outcome in the rescue effort.

In many ways there is an interesting parallel between risk in the rescue business and risk in the leadership business. Just like risk in the rescue business, leadership in an organizational context involves the lives of people and the potential death of the organization if risky decisions are not properly vetted.

Organizational leadership involves a risk-benefit evaluation because everything the leader does is based on the risk of the unknown. Taking people to a place they cannot yet see, or for that matter see the value in going to, requires great courage on the part of the leader. This is because, in carrying out this courageous effort, you are constantly battling both your own confidence in the potential successful outcome and those folks in the organization who may be averse to change. But when a situation reaches the top leader's desk, all the easy decisions have been made.

> Organizational leadership involves a risk-benefit evaluation because everything the leader does is based on the risk of the unknown. Taking people to a place they cannot yet see, or for that matter see the value in going to, requires great courage on the part of the leader.

So let us examine leadership from the risk-benefit perspective. If you look at a risk scale along a horizontal plane with least risk on the far left and greatest risk on the far right, neutral risk falls somewhere on the midpoint of the line. Those decisions with the most information regarding the situation and potential outcome are to the left of center, and those with the least information are on the right side.

Just like employees who deal with change, we as leaders find comfort in being located somewhere on this line (fig. 8–1). We may have tendencies that move us to different places on the line depending on the type of issue, previous experience,

Leadership & Risk

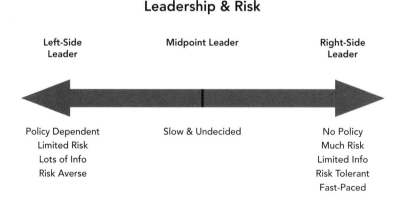

Left-Side Leader	Midpoint Leader	Right-Side Leader
Policy Dependent	Slow & Undecided	No Policy
Limited Risk		Much Risk
Lots of Info		Limited Info
Risk Averse		Risk Tolerant
		Fast-Paced

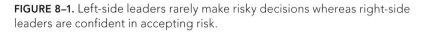

FIGURE 8–1. Left-side leaders rarely make risky decisions whereas right-side leaders are confident in accepting risk.

or just plain confidence in handling negligible risk. As leaders we develop techniques to handle situations and find comfort regarding outcomes.

We move along the line but nonetheless will settle within a certain range that seems to have produced previous success. Some employees are resistant to change while others remain ambivalent to the new course of action. Of course, we also can't forget about those who are excited and open to change. Using this analogy, leaders can also be located somewhere on this line when it comes to leadership and deciding action steps. The propensity to be in any one place on the line is dependent upon the information available and the amount of risk they are willing to take in the absence of a full, 20/20 peek into the future.

Left-Side Leaders

To see where leaders might fall on this risk-benefit line, let us first examine a left-side leader. This is the leader that first looks at a situation from the "me" perspective. In a nutshell, they first ask how the potential decision will affect them personally as a human being or professionally as a leader. This kind of leader needs to be clear as to what is required for a successful outcome if the important points of success are not already known.

When you sit down in front of a left-side leader, they will listen intently as you describe the situation. This great listening skill is honed from years of trying to ascertain clarity to the point they can see the final picture. They cannot and will not get behind anything that is not a slam dunk idea with minimal potential for organizational mayhem.

Let's imagine a scenario. Brian is a left-side leader. Julie, a human resources technician, has come to Brian because she wants to recommend a new compensation strategy that is geared toward employee retention. She fears that because recent raises have not moved people in their pay ranges, they could well become a training ground for new employees that will eventually leave the organization.

Julie is a true leader. Her style is to seek out problems and address them before they create a crisis for the organization. Because she solves problems before they occur, her efforts are rarely recognized for her worth to the organization.

As she lays out the details of her plan, Brian is listening carefully; however, instead of listening to the facts or challenging her strategy with inquiring questions, he is thinking about how the suggestion will be viewed by the board of directors. Will this program be perceived as too expensive? Can he be certain that the new program will retain employees and not just cause additional payroll? Can he be certain the outcome will be positive? Is there any danger that he could be viewed as giving direction in a situation that will eventually net an unfavorable result?

As the meeting progresses, it becomes clear to Julie that Brian is not comfortable with the plan. His input is framed in a manner as to suggest that this situation certainly could be a problem if not addressed and that Julie should stay on top of it by continuing to examine trending data. When the trending demonstrates an actual problem, they can revisit the situation. The bottom line here is the left-side leader provides no commitment because risk and project success are unknown.

My suggestion for you regarding left-side leaders is to watch out. They frequently are so ambivalent when giving direction that followers have no clue what they are expected to accomplish. This kind of leader will try to give direction that can be framed as one way or the other. It is a self-preservation style that will leave you on the short end of the stick when it comes to support.

The left-side leader by nature appreciates crisis-management leadership more than problem-solving leadership. A crisis can be seen and is clear. Something has happened, and something must be done. In a crisis, our elected leaders and board of directors are at the very least more tolerant of bad decisions on the part of the leader. Crisis leadership is the easiest of all leadership because, in general, people are more forgiving when balancing the speed at which the decision is made and the speed at which the situation is unfolding.

Do you think by not taking any risk, the left-side leader is taking any less of a risky position? If the left-side leader is risk averse and the organization effectively fails to prepare for the future, is that not risky? If such a leader lasts a long time in an organization, however, and fails to move toward success, how will that affect their legacy?

Midpoint Leaders

Unlike the left-side leader, the midpoint leader is only partially paralyzed with fear. They appear fearless because they know from the outset that they are not going to take a chance. When a decision is reached, it will be when the situation is either on the way to being a crisis or it has become clear enough that everyone can see what the next move is supposed to be.

The midpoint leader may at times appear to be a paralyzed leader. This is because they are the leaders that allow the world to come to them, and when direction and outcome are known they decide. Often, this type of leader will not make many mistakes; however, they also never really do anything but keep the organization from sinking. This is sometimes called damage-control leadership.

In another scenario, John is a recently promoted engine captain. He is smart and has managed to have a career that is free of mistakes because he is a black-and-white kind of thinker. He knows the policies of the organization by heart and adheres to them to the point that his subordinates are afraid to make any decisions outside of known policy.

On the way to a house fire, John realizes that he is going to be the first to arrive on the scene. The report is that people may be trapped, so there is urgency about getting to the scene, establishing an initial attack line, and completing a search of the house.

As the engine is approaching the last hydrant on the dead-end street, the driver asks John if he wants to stop at the hydrant and lay in a supply line. John knows that the department policy specifies that the first engine should lay a line on dead-end streets because if they don't, apparatus arriving later may not be able to get a line established before the initial arriving unit runs out of water. In addition, the next-arriving units may need to back down the street and lay out to the water supply.

"Captain, you want to hit the plug?" Silence. "Captain, I am getting ready to pass the last hydrant, do you want me to stop and lay a line?" Silence is all that remains as John and his engine pass the hydrant and continue toward the house fire.

What causes John to lock up with fear and become unable to decide? This decision to enact an aggressive attack and rescue should have been an obvious choice. Policy says to lay a line down dead-end streets; however, the element of timeliness was introduced into this equation because all involved knew there was the potential that people were trapped. John's problem here is that he is a midpoint leader when it comes to risk. John will deviate to some degree on either side of no risk and risk, but do not expect him to go too far on one side or the other. That just is not in his DNA.

One thing that you can absolutely expect is that the midpoint leader will always teeter on either side of the midpoint. They are willing to take some risk but never get carried away. They will seek the protective council of policy because they know that in doing so, they can never be held accountable for making a decision that is too far outside of policy.

You can recognize a midpoint leader in meetings because they miss the leadership moment, the moment when all others are looking for the leader to say, "I have heard enough, this is what we are going to do." If you find yourself walking out the door of a meeting even though nothing seems to have been accomplished, then you have just been in a meeting with a midpoint leader.

By nature, John will most likely wait until the situation is tilted toward the no-risk side of the scale to act. At the very least he will seek to find balance on the continuum. Midpoint leaders like John need a safe place from which to operate; in this case, there was ample reason to deviate from policy, but because the risk of self-preservation outweighed the risk of making a decision outside of policy, no action was taken.

The problem with midpoint leaders is not that they will make a crazy decision. It is that they can seldom decide in the time it takes to keep the organization on the front side of innovation and preparation. That is where the right side of risk lives and prevails.

John will have a great career in most cases, as he will not make any mistakes and will do well at handling a crisis when it occurs. But this is only because he will have the help of others to manage crises and he alone will not be held accountable for the results.

Right-Side Leaders

The right-side leader is an uncomfortable but highly confident leader. The emphasis is on confidence because they would not be able to advocate for organizational changes that prepare to move the organization forward without being able to step up and say, "This is the direction we are taking, and we are taking it because I believe it will best position the organization for future success."

The right-side leader is always taking the organization to a new place. In many cases, this is a place that others in the organization cannot see or, even more challenging, do not see any reason to go to. They spend lots of time convincing the organization that the direction is appropriate, as they fight both their internal confidence and those in the organization who are change averse.

Josh is a right-side leader. Josh understands that there is risk to being an effective leader because decisions regarding his direction will not be evident until sometime in the future. Even more disheartening for Josh is that his critics will be long ago silent as success finally reaches the organization.

Josh, in a meeting with his left-side boss, advocates for a change in policy regarding time and attendance recording for lunch breaks. It seems that in the current system, some employees are clocking back in a few minutes early primarily because they are through eating and want to get back to work and do a good job.

The problem is that the current system, while accurate and dependable, makes employees feel they are not trusted because they must clock in and out on the half hour in order to satisfy the Fair Labor Standards Act rule that employees must receive a minimum one half hour uninterrupted lunch break each work day.

Josh recommends to his boss that the organization could direct the clock to adjust the half hour automatically and then establish a policy that puts the accountability for compliance on the employee to manage their time under the rules established by the Fair Labor Standards Act. Josh's idea is a simple one: treat people like adults and then expect adult behavior in return. Josh's boss has little faith in the organization's managers to operate in this type of environment.

The problem, as defined in this case study, is a contrast in risk versus reward. If managers do not hold their people accountable, the organization could be held legally liable for the infraction. From Josh's point of view, the reward of employee engagement and the message of accountability and trust that is sent far outweigh the risks associated with a potential violation by employees.

Josh's problem is that he is working on an issue that is not evident to the boss. He is working in an area that is intangible at best. For instance, should he be right and the organization's productivity improve, how will it ever be attributed to this decision? On the reverse, Josh's boss only sees risk because he is a left-side leader. The intangible potential outcome is not enough to warrant a risk to change the policy.

Josh becomes frustrated because he understands that true leadership occurs when the leader not only balances risk and benefit but also is not afraid to operate on the right side. Sure, there is a chance that things will not happen exactly like he planned, but what if they do? What if employees see themselves as empowered to make their own decisions and this empowerment leads to increased productivity?

> Everything about leading organizations is in some way or another related to managing risk when it comes to the future.

Frequently, right-side leaders are not respected for their leadership until they have long since left the organization. That is when someone will say, "Man, I am glad that dude was here back then because he laid the groundwork for all of this success."

The point to all of this is to ask yourself whether you are stuck somewhere on the line or moving along the line based on the circumstances presented. Just like in the rescue business, we want to move along the risk line, as each decision point is based on a complete assessment of the risk compared with the benefit. Effective leaders move along the line when they are making decisions about an organization's future. Successful leaders do not take unnecessary risk, but they do take risks. Everything about leading organizations is in some way or another related to managing risk when it comes to the future.

So the question becomes, what kind of leader are you? Are you risk averse unless you have all the information? Or are you paralyzed with fear and hang in the middle where policy and outcome can be determined? Or are you the kind of leader who is brave enough to evaluate risk and confident enough to make decisions in the absence of absolutes?

I hope you will do two things with this new information. First, recognize if your tendency is to be in a certain place on the line regardless of the circumstances. If this is the case, ask yourself if this tendency is helping your organization position itself for the future or rather just keeping it afloat. Second, recognize that true leadership almost always occurs on the right side of the line. In this case, being on the right side of the line does not mean you jump off a building and hope you are the first human being who can fly without wings. But it does mean it will take courage and you will have to take risks to be an effective leader. The difference in successful and not-so-successful leaders will always reside in the success they have when managing the risks on the right side.

All leaders operate at many different points on the scale depending on the complexity of the situation and the potential consequences of the outcome. High performance leaders evaluate every situation for the risk versus reward and then shift their decision-making based on the information available and the immediacy of required action.

9

Leadership and Organizational Balance

Open and honest dialogue, and the willingness of the leader to share leadership responsibilities, is the key to organizational balance.

Key Points

- Great organizations have good organizational balance.
- High performance organizations seek compromise to maintain balance.
- Great leaders surround themselves with people who bring balance to their style.
- Leadership in a healthy organization shifts but always maintains balance.
- Too much conflicting leadership in an organization can cause overpressure.

I have come to believe the world must be in some state of balance for all things to be in good order. In my trench rescue work, I teach students to understand the art of stabilization from the point of view that the earth is always trying to be in some state of equilibrium. When we put a hole in the ground, we upset this equilibrium, and as the hole gets deeper, it keeps getting further and further out of whack.

In the case of trench stabilization, if we intercede before the hole gets too deep and out of balance, we are successful. If the hole gets too deep before we intercede, the earth takes care of the situation itself by collapsing and therefore bringing things back into balance.

When it comes to nature and its affinity for being balanced, we try to trick the earth into thinking it is in balance and all pressures are equal. We accomplish this using shoring that transfers the energy from one side of the trench to the other and bridges the gap. Ultimately all this effort brings balance to the situation.

To prevent a collapse, I have always tried to seek balance in my personal life and do my best to close the gap when things are out of balance. This sometimes means tempering my own behavior and at other times it means asking another party to temper theirs. In most cases, seeking compromises that narrow the gap brings balance and is all the shoring necessary to prevent relationship collapse.

If you need metaphors to understand the principle of balance, think about a team of runners. Not only do they all want to go fast but they all want to beat each other. Soon all they think about is going fast, and speed becomes the yardstick by which the group measures success—not whether the root cause of the issue is addressed or if all the stakeholders are considered, just how fast they come to doing something.

In this example someone in the group will win because they run the fastest. For this group of people, the problem will always be that the winner will be an individual and not the team.

So how does balance help this situation and make sure the group is running the right race at the right track? Certainly, a slow guy on the team would encourage the group to stop and catch their collective breath at some point in time. If someone who was not a racer at all was involved, perhaps they could tell the team that they were at the wrong track in the first place.

The relationship/sports balance problem also finds its way into our work life. Work groups and teams that have too many people of one type or the other soon find themselves out of balance. If all leadership is left to business folks, the group may be viewed as only caring about results and not concerned with the people who work for them. Likewise, if too many on the team are overly concerned with operations, they could be called micromanagers.

The lesson for leaders in this example is to seek ways to bring balance to the work group, team, or organization. The leader should understand that, to maintain balance, a situation of opposing forces that push and pull needs to be created.

How Does Balance Relate to Leadership?

It is my belief that people are replaced in jobs because they get too far out of balance one way or the other: too hard or too soft, too compassionate or too callous, or too much a visionary at the expense of handling their daily work or too focused on the present to plan for the future. The examples of potential opposites in behavior and style are endless.

Take a close look at what type of person follows someone into a job as an appointed replacement. They will likely be the opposite of the person who left. In most cases the incumbent will be perceived one way and the organization will be convinced that all things will improve if they hire someone who is opposite.

This is an example of an organization trying to bring about balance too quickly after the first leader leans so far one way that there becomes desperate need on the part of the organization to course correct. Success in this situation will only come when the new leader reaches the point where their style balances the organization. In time, if they are truly opposite to their predecessor, the situation will present itself again, just in the other direction.

This type of situation happens so often that if you are applying for a position, you will do well to profile the person you are following. If you are the opposite, then you have a good chance of landing the job. Just keep in mind that success will be

> Long-term incumbents in any leadership position find a way to create balance. Either this is accomplished in modification of their own styles over time or by recognizing the wisdom of surrounding themselves with people that bring balance to their leadership style and team.

fleeting because you too may find yourself on the wrong end of balance one day.

Long-term incumbents in any leadership position find a way to create balance. This is accomplished either in modification of their own styles over time or by recognizing the wisdom of surrounding themselves with people who bring balance to their leadership style and team.

Now let us examine the balance issue as it applies to organizational leadership. In this case, let's look at the balance of leadership in an organization as if it were a balloon, which has balance in that the pressure inside is the same on all the interior walls.

Imagine this balloon as filled up full of organizational leadership. At any given time, the same amount of leadership always exists in the balloon, and this is what helps it maintain its shape, pressure, and size. Formal and informal leaders up and down the chain of command in your organization, as well as the employees' union, are a part of what makes up the leadership inside of the balloon.

So how do we keep the balloon in balance and shape? How do we keep the balloon from being overpressured and bursting because of too much leadership? Understanding this dynamic as a balloon is a great foundation for us to examine how leaders use leadership in successful organizations.

Leadership is necessary. Do not buy into the fact that any team or organization is leaderless. You hear this sometimes when an organization is suffering through a poor leader, but believe me, someone or something is going to fill the leadership void when the formal leader does not lead. The bottom line is that the balloon will stay full of leadership regardless of who the appointed formal leader might be.

The formal leader, like the chief, should control a large volume of the leadership contained in the balloon. Just how much shifts from time to time is determined by what is occurring in the organization at any given time. When the formal leader leads in proper balance with the other leadership in the organization, there is mutual respect for all levels of leadership and what they bring to the table. In effect, the pressure in the balloon stabilizes.

Where leaders frequently go wrong is thinking that their title dictates the percentage of the balloon they have to themselves. The title in and of itself only provides credibility when the leader walks the talk and the other leadership elements in the organization respect them for it.

Leaders who find themselves under pressure and withdraw to seek safe refuge must understand that other elements of the organization push in to fill the void. In some cases, these elements are good and positive, but in other cases, they can be distracters that are looking for any measure of power in the organization. Make sure you understand that withdrawing to conserve power just gives it away. Holding on to anything without investing in it ultimately diminishes its value.

For example, let us consider the chief who loses their ability to lead because they violated the closely held values of the organization. Recognizing this void, the union president starts dictating overall direction of the organization and becomes a more powerful influence in department direction than the chief.

Now before any union folks get their feelings hurt, I do not intend to imply that they are not a part of what fills the balloon; they just should not have the most volume in the balloon at any given time. They certainly need to be in there and for good reason.

Smart chiefs understand how the role of unions and that of other formal and informal leaders in the balloon are necessary. We are always striving toward balance. Every smart chief knows their limits when it comes to politics. For instance, every city or county manager recognizes that the fire chief is somewhat of a community figurehead and, in some cases, carries tremendous community support. These managers should and most likely will spend some time with the chief indicating their tolerance for political activity and subsequent dialogue with elected officials. Whatever you do, understand the limits and respect them.

There are times when the fire chief wants to influence a political decision, but because of the bureaucratic and political boundaries of separation, they are unable to stand up and voice their opinion. These are times when the chief, having a good relationship with the union, may want to pull back a bit and let them politically advocate a desired position. This recognizes another type of leadership in the organization and uses it to some sort of successful advantage. A word of caution here is appropriate; always make sure your boss is aware of these situations and in all cases make sure your behavior remains credible and ethical.

In still other cases, it may be as simple as the fire chief needing buy-in from the members on a new initiative and therefore allowing several senior employees to influence department direction. It could also be a situation when the chief may want to let peer pressure correct an undesirable situation at one of their stations. Whatever the case, recognizing these subtle shifts in leadership and maintaining leadership balance in the balloon is vital if the chief wants to be successful.

Lastly, the chief should recognize that the balloon can burst from overpressure created by too much competing, out-of-balance leadership. If all the leaders in the balloon want the same amount of volume, the subsequent pressure will surely cause the balloon to burst.

Having open and honest discussions with your organization's formal and informal leaders is key to a leader's success in maintaining leadership balance. This is where the value of good relationships with your employees lends itself to effective leadership. The employees will know that in some cases you are willing to give up power and share it with others in the organi-

> Having open and honest discussions with your organizations formal and informal leaders is key to a leader's success in maintaining leadership balance.

zation if this is what maintains leadership balance and takes the organization toward its desired goals. They will also know that you are comfortable in your role as leader because the message you send to the organization is that you recognize everyone is a leader and you are willing to share that responsibility with them from time to time.

One of the most important elements in organizational leadership is grounded in the balance you build into the environment in which you operate. Maintain balance and you will maintain progress and success that comes from forward motion. Do not be afraid to share leadership responsibility when necessary, and in all cases recognize that maintaining consistent leadership in the balloon is necessary for success. The trick is to always know what part of the balloon you want to maintain and when some shift is necessary to achieve success. Remember our organization, just like the universe, is always seeking balance and will find it with or without your participation.

All things in life require balance and equilibrium. The high performance leader understands their tendencies, or behaviors that are prevalent with the leaders in most situations, and then strategically places people around them who bring about balance in those tendencies. Imbalance in one direction or the other, while effective in some situations, may not be in others. A healthy respect for all employees being leaders brings about organizational balance as the leader and organization seeks equilibrium.

Conflict Resolution

Just like a hurricane, it is always best to deal with disagreeable people after they have blown all the wind out of their system.

Key Points

- Stay calm when faced with adversity.
- Listen carefully for the real nature of the conflict.
- Spin bad situations into positive outcomes.
- Winning a disagreement might not really be winning.
- Resist the urge to get emotional.

I wrote this chapter after returning from a Federal Emergency Management Agency (FEMA) deployment for Hurricanes Frances and Ivan in Florida.

During the 18 days I spent in Florida, I had the opportunity to work with some of the most professional and expert rescue people in the world. I love the men and women in Urban Search and Rescue (US&R) because they are outstanding rescue people and they teach me so much. People like these, and others from all walks of life, seem to come into our lives and then leave having given us the best (or worst) they have to offer: the best via their operation under stressful situations and the worst when they have too much down time. In that respect, US&R is no different from being assigned to a fire station.

During the Ivan part of the deployment, I had the opportunity to meet President Bush, along with his brother Jeb, who was the governor of Florida. I found both to be very caring and compassionate people when it came to interacting with the rescue personnel who were present and the citizens, many of whom had lost everything they owned.

So you would think that I would spend my time writing this piece to speak about that experience, but quite frankly, that is not the one experience that I learned the most from. The man who taught me the most on this occasion was the person who treated me the worst.

Sheriff McNasty

As one of our long days was winding down, I heard ranting and raving coming from the opposite end of the command trailer. I was on the phone with the Joint Management Team in Orlando, so I was trying not to pay attention to this unwanted intrusion. After a few minutes I was advised that the local sheriff wanted to talk to whoever was in charge, and his intention was to give this person a piece of his mind.

Imagine the sheriff from *Smokey and the Bandit*, only slightly more polished and even meaner, standing in front of you pointing his finger and yelling about how you are not authorized to be here, how he is not informed on what you are doing, and how he could send all of us off his property if he desired to do so!

Hurricane Ivan was only a light breeze compared with the storm surge and wind that had come out of that guy once he was finished with me. To add insult to injury, I had to face his county manager and an arrogant second-in-command deputy, who both felt inclined to pipe up every time the sheriff took a breath.

When his tirade was all over, I managed to finally figure out that the sheriff had just been on television explaining why a barrier island was not being searched while the background footage they were showing was of our folks being dropped off by helicopters to search that very island. To say he was embarrassed would be an understatement!

In the back of my head, I was wondering why this guy was so uninformed since we had been fully operational for two days in full cooperation with the local command, a local command that incidentally included one of his deputies, who was obviously not keeping his boss informed of our planning meetings and rescue operation results.

I did manage to smooth things over with the sheriff and told him that I would personally brief him daily as to our activities in his county. He seemed somewhat placated by this gesture and the three of them left us in peace to continue our operation.

When the trailer cleared, my division supervisor, who was a battalion chief from Menlo Park, California, asked if I was aware of the sheriff's deputy being involved in all our previous briefings, and I said yes. His next question was, "Why didn't you just tell him so he would shut up?"

Believe me when I say that I have dealt with a lot of mad people before, and none of them want to be told they are wrong when they are ranting and raving about being wrong in the first place! Therefore, I told the battalion chief that the important thing was that we started the process of turning the situation around. Besides, the sheriff just wanted somebody to blame for being uninformed and looking like a doofus on television.

The next day, the Florida state coordinator for US&R, my division supervisor, and I went to the Escambia County Emergency Operations Center (EOC) for a briefing for the county command staff. On the way there we noticed a voting sign for this sheriff, who just happened to be running for reelection. "Wouldn't it be funny to get a picture of us by his sign with our hands in a thumbs up position?" someone said. I was thinking about a different hand gesture, but common sense prevailed, and we took the originally proposed picture.

> Believe me when I say that I have dealt with a lot of mad people before, and none of them want to be told they are wrong when they are ranting and raving about being wrong in the first place!

When we finished our EOC briefing, the EOC coordinator told us that the sheriff was known for his short temper and that we should not feel too bad about our encounter with him. I thanked her for the advice and told her about the picture we wanted to give him as a good faith gesture and our commitment to being on his team. To this she replied, "I will see that the sheriff receives the picture."

Little did we know that the sheriff would receive his picture at the EOC briefing that night in front of about 50 people. While most of them laughed, I am told the sheriff was once again embarrassed, and my friends said I was now a marked man. I fully expected him to run out from behind the bushes the next day and let me have it.

As it was, I needed to meet with the Secret Service to prepare for the president's visit. (I will save that story for another time but suffice to say I did not have time to worry about the sheriff.)

I received explicit instructions as to how to greet the president, where he would be dropped off, and the route his motorcade would take through the neighborhood, even so specific as to where to park the bus the other rescue folks were riding on.

As we approached the president's arrival point, I got out of the bus to organize the parking, and you will never guess who pulled right in the place where the bus was supposed to be—that is right, the mean sheriff! In all his glory, pomp, and circumstance, he was right where I needed to park.

I was not looking forward to how this conversation was going to go, but nonetheless I approached the sheriff and told him I needed to park the bus where he was parked. To this he replied that moving his car would not be necessary because he was clearing the area. He further stated that the bus, all the folks I had with me, and I would be moving out of the area because the president was coming. I replied that I knew that, however he cut me off with another of those "Don't you remember who I am?" sermons before I could explain my direction from the Secret Service.

I calmly walked away and searched for my new friends in the Secret Service to tell them of my dilemma. By this time, the Advance Team had transferred command to the Ground Team. The Secret Service Ground Team guy just happened to be the biggest and most serious guy I had ever met, and he was not happy that the sheriff was messing up his plans for the president's arrival.

The three of us walked very hurriedly over to the sheriff, and my new big friend told him to move his car or he would arrest him and place him in one of his own squad cars. Furthermore, he noted he did not have time for this as it was almost "game on." Somehow the old sheriff lost a little of his wind, and with his head down started to move his car.

Suddenly, the Ground Team leader told the sheriff to drive us to the president's point of arrival. We all jumped in the sheriff's car and headed down the street. As he stopped the car and we got out to get into position, the sheriff asked the Secret Service agent where he wanted him to park his car. To this the agent replied, "Back where you just left from, and hurry up about it." As I exited the car, I very politely thanked the sheriff for the ride and calmly gave him a thumbs up. At that point I had said all that was needed, and I am sure the sheriff left wondering whether it was a genuine thank you or a sarcastic recognition of how he had treated me. The answer to that will remain solely with me, and I will leave it to your imagination to decide for yourself.

The president did arrive, and I was honored to greet him and explain what his rescue teams had accomplished in Florida. He appeared a very caring, concerned, and compassionate person. His questions were good ones (figs. 10–1 and 10–2).

Here are some suggestions when you find yourself between the sheriff and the door:

When someone is treating you unfairly and yelling at you like a wild man, it is seldom good practice to confront them. Let them vent, and then try to determine the root cause of their frustration. Often, it is something other than what they are saying to you. Asking someone "Why?" five times is a good technique to get to the root cause of the issues.

Listen very carefully and do not interrupt the irritated person. Let them go on until they are all out of wind.

FIGURES 10–1 and 10–2. A meeting with U.S. President George Bush in Florida (courtesy of Jocelyn Augustino).

Look for some point in the conversation to turn the whole thing into a positive exchange. This is where you may need to eat a little humble pie.

Understand that winning the argument is not what this is about. The focus should be on what is required to get something positive accomplished in your mission.

Never say anything personal about the angry person or be rude back to them. You should resist the temptation to get emotional. Remember, in most cases you are not the one they are mad with and much of the show is about nothing more than allowing them to vent and blow smoke.

If someone is yelling at you, it is because they cannot find a legitimate way to make their point. Believe it or not, people who conduct themselves in this fashion provide you with the advantage because they are letting emotion drive their words and are not thinking about what they are saying.

In the end, understand that life tends to work these things out. Mean folks who demean others and have their own self-interest in mind generally get what they deserve, even if you are not there to see it for yourself. Just wait long enough, and you will get your chance to give them a thumbs up.

When dealing with an adversarial situation, the person who remains calm is always in a position of power. Emotional people are rarely rational in their thoughts, as their mouths work faster than their brains.

As a tactic when dealing with unreasonable people, always listen carefully to what is being said, but never assume that what is being said is the real problem. Anger manifests itself in a mostly personal fashion and for the most part is devoid of any attention to address the actual issue.

Regardless of a difficult confrontation, always look for a way to create at least some sort of positive outcome. Progress can be made when there is something positive to look forward to; however, only further disagreement will result should you part ways angry and negative.

In dealing with difficult people, recognize that you are never in a position of power when concentrating on winning. Winning is not the answer, as that by definition infers that someone is losing in the situation. Resist the emotional aspect of confrontation by seeking compromise and therefore positive progress at resolution.

11

Valuing All Employees

*Every single day you lead is an interview day when people
are formulating what will be your leadership legacy.*

Key Points

- Value all employees.
- Listen to your employees.
- Beware of shifting momentum.
- Convey an appreciation for timing.
- Be inclusive.
- Look and think long range.
- Do not allow employees to disengage.
- Create a mentorship culture.

When comparing fire and EMS trends, there are many consistencies in what we experience organizationally on a daily, monthly, and yearly basis. Funding issues, employee health and safety, staffing, equipment replacement, and capital replacement are just a few of the complexities all of us deal with on some level.

Whether in a volunteer or a career organization, the situation we are all experiencing, or anticipating, is the mass exodus of our senior employees and influx of new, inexperienced employees. One reason for this phenomenon is the rapid expansion of the career fire service during the last several decades. The seasoned firefighters are now leaving and taking with them their tremendous wealth of experience and institutional knowledge.

For many of us, this wholesale change has already occurred, and we are dealing with the challenges created by the turnover. For others, these issues are just rearing their ugly heads, and because of slow planning, we are struggling to introduce innovative programs to minimize the impact created by the "experience vacuum."

It is the leader's responsibility to prepare for and recover from these changes in our organization. One place to start is to create an environment where senior, nonretiring current employees can continue to prosper and flourish while new employees are being integrated into the organization. This would make all employees feel valued and would facilitate this transition by transferring the wealth of organizational knowledge and experience carried by our senior employees to our new employees.

The wholesale replacement of personnel presents enormous challenges for leaders. Lack of on-the-job fire experience for new members is a paramount concern, but there are others as well. Ultimately, the department's success may depend on how skillful the leader is at being able to keep all department members engaged in the organization's work and in establishing mechanisms for transferring the knowledge, skills, and abilities of the senior members to the less experienced ones.

In the last century America was burning. Many of you hired during this period can remember running multiple fires in one duty shift. Multiple-alarm fires and fires in general were frequent enough to allow recruits to learn on the job. Recruit classes were a few weeks shorter because organizations could count on their seasoned veterans to teach new recruits the tricks of the trade while on the job—literally on the fireground!

Those times bring back fond memories of fighting fire for the first time and having a crusty old veteran firefighter tell me, "If you don't get your head out of my butt while we crawl down the hall, all you will ever learn is what my backside looks like."

Some of the folks we are promoting these days do not have very much experience. Some of them may have only a few fires under their belt, and their nozzle time may be limited to just a few room-and-content fires. In many cases, these newly promoted employees are getting experience at the same time as the recruits.

I was speaking with a battalion chief from the New York City Fire Department (FDNY) concerning the effects of retirement on its workforce. You can imagine the problems created in such a large, active department when those who are leaving are not only the older employees but also so many other members who have been affected by the World Trade Center attacks. (In addition, numerous experienced veterans were among the 343 department members killed in the collapse of the World Trade Center towers.) The chief said, "I hear more Maydays now than ever before. They are for the most part being transmitted by newer, less experienced officers and firefighters who in many cases don't have the experience the older firefighters had."

Not only are our senior folks leaving us, but many of them are also distraught and angry when they retire. Sometimes these emotions are self-induced to create a justification for leaving one stage of life and heading to another. I refer to this response as a self-preservation instinct to justify a change in their own mind. However, many of the senior employees are distraught because leaders have failed to demonstrate to them their value to the organization.

A good example of this is when we give developmental projects to younger officers because we desperately want to give them confidence and experience in managing people. All the while, we know full well that many of our experienced folks could do these things standing on their heads. The question I want to ask at this point is, are your methods inadvertently excluding your most senior and experienced employees? You know as well as I do that as soon as something goes wrong, we expect these senior employees to bail us out by coming to the rescue.

We must use this strategy to develop people. The problem may be in the message and how we deliver it. These situations beg the question, are we engaging our senior employees in helping to train and mentor our younger personnel or in leading important departmental projects? What does it say to our senior folks when it appears that they are not valued until we get in a bind?

> If your organization is not already including senior employees in departmental activities, or if there tends to be an attitude that senior folks are tired and just riding out their careers, you may have already lost some of these folks.

If your organization is not already including senior employees in departmental activities or if there tends to be an attitude that senior folks are tired and just riding out their careers, you may have already lost some of these folks. Typically, the behavior associated with this withdrawal is trying to just get by, not volunteering to participate in the work of the organization or simply saying no to every request, using excuses such as "I don't have the time" or "I'm way overextended now." Worst yet, some of these folks may sabotage current change efforts so they feel more comfortable. Remember, most employees develop a perception concerning the way things are going to be when they retire or leave the organization and for this reason resent efforts that change this paradigm.

Another very important consideration is that some of our senior employees may not have the skill set required to maintain efficiency in today's modern times. For instance, if they are technologically challenged or cannot quite do the math to be a hazardous materials technician, this does not make them stupid or less desirable employees. Nor does it mean they cannot add value to other departmental initiatives. It just means that we as leaders did not at any point create an environment in which these employees were motivated to learn and change.

So, what can we do about these complex issues involving our new and senior employees? What can we do to engage senior employees and not inadvertently exclude them from the department's work while helping new employees to grow?

We can all admit that when times get tight, we have our go-to folks, the people who always come through when it is really on the line. Our organizations are full of people with varying levels of skill. Some have confidence; some do not. Some are very intelligent, and some have just enough cognitive ability to pass the civil service exam and understand hydraulics. The leader needs to figure out how to engage all employees in ways that they can add value and stand a good chance at success. Above all, make sure you value all your employees and especially the senior employees you have been counting on all these years.

Consider all stakeholders, respect job descriptions, and match the job or task that needs to be accomplished with the roles and responsibilities already outlined in the job description or organizational structure. This will go a long way toward valuing every individual for the job they are paid to perform.

As a leader, being told to listen may sound a bit trite. As we all know, listening is a required leadership skill. The problem is that sometimes we listen but do not hear what our people are saying. There may be a tendency to just listen to the complaint (symptom) and not hear the real problem (disease).

> It is imperative that leaders realize when the pendulum has swung so far toward the growing of new employees that senior employees are excluded.

Failing to determine the root cause of problems when we listen to employee concerns ultimately just placates them. Believe me, all employees want to be heard, even when they do not or cannot divulge the real problem. It is our responsibility to figure out the real problem if it is to be solved and the employees are to feel you value what they have to say.

It is imperative that leaders realize when the pendulum has swung so far toward the growing of new employees that senior employees are excluded. This is another case where including senior people as coaches or mentors demonstrates a commitment to what they bring to the table in experience, even if they ultimately know the result will be someone else getting the benefit. It should never be acceptable to deny senior staff opportunities to impart their knowledge, skills, and experience before leaving. That is the only way to keep them involved.

When developing project teams, make sure you include all employees, with a good cross section across all ranks and levels of experience. This type of inclusiveness can benefit the organization in many ways, especially in capturing the experience of your senior employees while they are informally mentoring the newer ones.

This does not mean that the most experienced person should always be assigned the formal leadership position. It very well could be that you are trying to get a young officer some experience and that interacting with senior folks in an informal/formal work environment might be beneficial. The optimal outcome is a successfully completed project by employees who have grown because of their participation.

Begin with the end in mind and think backward about assignments to balance the knowledge, skills, abilities, and competencies across multiple projects. Post the plan for the year (or 18 months) and share what you know and when adjustments must be made. Tell members why this or that individual was selected to participate in a given project. Pick the level of position appropriate for the assignment. If it is strategic in nature, select more experienced staff and add a few new folks for mentoring and coaching.

> Process improvement and change are about customer service and how to create environments wherein our people flourish and provide more efficient service.

I put stock in the saying, "If you work here, you have a responsibility." That responsibility transcends all aspects of our work and service to the public. The leader's job is to create an environment in which all employees are engaged in the department's work.

An effective leader in a successful organization is committed to building a culture that continually mentors and grows all employees—a culture that makes sure all employees understand that if they work here, they are expected to continue to grow and evolve. This means that all members understand and accept that things will change and there is no guarantee that the organization will remain a certain way during any portion of their career.

This is especially important in today's workplace because the delivery of fire and EMS services is more and more based on technology that is evolving at record speeds. Our personnel need to grow accustomed to change and its effects.

As the leader, you must stay engaged and realize that you are responsible for your organization's future. You determine whether your department's culture will be one in which learning is encouraged or change is resisted. You cannot delegate this function to the training division or expect that it will go away because you look away from it.

People will remember the legacy we leave as leaders. This is as true of our first assignment as an officer as it will be of our final leadership position. As leaders we are stewards of our organizations. We exist only to guide, mentor, develop, and align our employees' efforts. Understanding our role and accepting the responsibility to further our organization's capability while keeping all our personnel engaged is hard work, as you already most likely know.

Most organizations are in a constant state of flux as older employees retire and younger employees come in to take their place. In a high performance organization, the leadership manages the fluctuations and changes in experience and talent by making certain succession planning and experience are transferred from one group of employees to the next.

Fires are less frequent than in the past, so real fireground experience is harder to come by. Making sure you engage your more experienced employees in promoting the transfer of knowledge is critical to maintain high service levels. Recognizing that all employees have a skill set from which the organization can benefit and then creating a mentorship culture is critical to leadership success in high performing organizations.

Part III

Organizational Leadership: Creating the Right Environment

The leader of a high performance organization needs to cultivate a facilitative environment and resist the pressure to lead by regulation and policy. You do not want your employees acting entirely because you have rules or boundaries in place. The much more successful approach is to create a culture where employees can act independently, or in concert with each other, to achieve the desired outcome of the organization.

Multiple Personalities and the Fire Service

Successful leadership happens when the leader can change the "us versus them" conversation to one that focuses on the many varied roles and responsibilities in an organization and not the conflicting priorities of individual members.

Key Points

- There will always be "us" and "them" in an organization.
- Appreciating all members' roles and responsibilities brings about understanding.
- Open and honest communication bridges gaps and changes the dialogue from individual priorities to team vision.

Lately, I have a hard time deciding if I am part of "us" or "them." You know what I am talking about, when firefighters are sitting around the kitchen table talking about administration, the senior staff, or me. In that situation I think I am a "them." Funnily though, sometimes when I am talking about line folks, they are my "them."

I guess what I am having a hard time trying to figure out is how we can all be an "us" or "them" but somehow remain different people. I know this sounds a little like the Abbott and Costello "Who's on First" comedy routine, but I guess that is the point. The problem is, it is not very funny and, in a way, is somewhat unhealthy for the same organization to refer to itself as "us" and "them." Let us look at a few of the problems this type of language causes the organization.

Organizations have been dealing with this situation since the beginning of time. However, there are a few things I think we can do to minimize the divisiveness that this type of talk brings to organizations.

There are many different jobs in a fire department, and all have very specific functions. Like players on a team, if one person does not fulfill their role, the success of the team may suffer. For instance, if the chief spends all their time on the line operating the system as one of "them" (from the perspective of leadership) then the chief will not be fulfilling their responsibility to provide vision and strategic direction for the organization. Likewise, chief officers need the "them," folks operating the system, to provide great service to our customers.

Look for the situations where we refer to others as "them" or to ourselves as "us." If we listen to each other and then question in what context this language is occurring, it will go a long way in helping us realize why it happens. Do the two parties differ in their view of a situation, or do they have different roles?

I believe that understanding our individual roles in the system and what value these roles bring to the table is critical. We need to be tolerant with each other and help each other learn about things we do not understand and what our individual responsibilities are.

Most departments have invested a tremendous amount of energy and resources into making sure they can communicate with each other. Standard operating procedures, guidelines, rules, regulations, roundtable meetings, newsletters, podcasts, Teams video, Zoom, and social media are all used for this purpose, to name just a few. The amount of time and money is all for naught if we do not make the effort to put good information into the system.

> Understanding our individual roles in the system and what value these roles bring to the table is critical. We need to be tolerant with each other and help each other learn about things we do not understand and what our individual responsibilities are.

Maybe the most important thing all of us can do is develop and maintain a constructive relationship with each other. Feeling free to express ourselves and to give and receive advice is an invaluable tool. It is the foundation established so that when we need support and help, we can ask for it without fear, retribution, or ridicule. The "us" and the "them" are the same folks after all.

As you begin, or continue, your journey into leadership, try to understand that leadership is not just one thing but includes many aspects. Leadership is of course about developing relationships with your people so that you know enough about them to take advantage of their best strengths. It is about trying to make sure you have the "right people, in the right place, with the right skills, at the right time."

Confident leaders understand that the "us versus them" conversations occur because there is not a general consideration for the importance of all roles in the organization. In this regard individual members, at all levels, may view their role as the most important and therefore they think of themselves as "us." Everyone who doesn't fit this paradigm in thinking becomes a "them."

It is useless for good leaders to fight the "us versus them" battle, and rather they should work diligently to educate members that these terms are most healthy when applied to imagining the perspective of the various roles in an organization and not when used as an example of individual conflicting priorities.

Setting the Stage

*The most successful organizational leaders are those
who have employees who thrive in an environment
of empowerment and system ownership.*

Key Points

- The paradox of leadership is that you are responsible for the delivery of services that are furthest from your control.
- Empowerment is the key to individual engagement.
- Holding power and control makes the leader less powerful and in control.
- Rules and regulations, if not essential to safety or organizational consistency, can signal mistrust and hamper innovation and self-motivation.
- Consistent decision-making and alignment of purpose comes from consideration of all stakeholder's input.

Can you imagine being in a more difficult position than that of being the person farthest away from the first point of service delivery and yet being ultimately responsible for it?

If you are a new chief officer, or even an experienced chief, it may be comforting to believe you are in full control of your organization, but this type of thinking will lead to a high-stress environment

> Can you imagine being in a more difficult position than that of being the person farthest away from the first point of service delivery and yet being ultimately responsible for it?

in which you will have a decreasing ability to cope with daily activities. In addition, you might end up with a department full of paralyzed workers who will not or cannot help you. Worst yet, the more involved you become in helping to correct the problems, the more ineffective your leadership will become.

Some leaders are good at letting the power go and leading their organization successfully. Others hold tight to the power because they fear losing control. Still others declare that personnel lack the experience or maturity to make system decisions. Sometimes, we hang our hat on a fire service culture that dictates we be involved in every aspect of our department's operation because "that's the way it has always been."

As mentioned in the previous chapter, when "us versus them" or "we and they" are frequently used to indicate frustration on the part of the employee, there remains a disconnect in understanding how individual roles, as different as they may be, all help drive organizational success. The point being is that you should strive to create a situation of trust where all employees do their part to help the organization achieve its priorities, mission, and vision.

As officers, we want the members of the organization to act and behave in a manner that brings credit to our department and the jurisdiction we serve. To accomplish this objective, we rely on standard operating procedures (SOPs) or operating guidelines that align decision-making and create a disciplined workforce.

Discipline, in this case, is not meant in the negative sense. It describes a situation in which the members of our team can be expected to perform in situations when they are given clear direction and, more importantly, when clear direction is not evident.

The problem with many of our SOPs is that they are usually written in response to some behavioral problem or even, heaven forbid, a single incident. Many are named after firefighter so-and-so, who behaved in some fashion we do not want repeated. The very nature of this type of response creates an environment of mistrust between the organization and its members by inferring they will usually make the wrong decision if not given definitive direction. In fact, it has been my experience that organizations with the greatest number of SOPs and written guidelines are least capable of managing an unsupervised workforce.

The nature of our business, and the necessity of command and control, dictates that we have multiple levels of command within our organizational structure. We are taught early in our careers about the benefits of span of control and unity of command. Those concepts get even more attention as we stress on-scene accountability for our personnel in emergency situations.

The problem remains that these levels of accountability do not go away during the 70% of the time we are not engaged in fireground operations. You could go so far as to say they sometimes bind our systems; create additional confusion between shifts, battalions, and divisions; and are directly responsible for inconsistent decision-making.

To move forward, you will need to buy into the belief that developing a platform for *consistent* decision-making will not only align your folk's decisions but may indeed lead to consistent employee behavior. With that little bit of an opening, you will discover that if a proper environment is established, the people in your organization will do the right things and consistent decision-making in your organization will become the norm not the exception, eliminating the need for an excessive number of SOPs and discipline.

For me, the term "strategic alignment" conjures up images of arrows all pointing in the same direction toward a stated and clearly defined outcome. This means that *all* people in the organization can understand, articulate, and visualize what needs to be done. Also, it ensures that at any given time all are "rowing the boat" to the same beat, the effort is consistently coordinated, and that available resources are ultimately maximized. Your need to be in control, therefore, will be diminished.

It is difficult for some chief officers to think strategically. If you are one of these folks, I understand. Getting caught up in the minutiae of daily delivery of our services and making certain that "moment of truth" service delivery is taking place is essential in some chiefs' minds. You can't argue with the fact that, ultimately, we are accountable for the service delivery and its related trials and tribulations.

To understand the challenge of a chief fire officer, you must first understand that our daily work lives are spent addressing the issue(s) of the day, answering a question for the city manager, dealing with a citizen's complaint, or mediating a dispute between employees, just to name a few. From a purely strategic standpoint these things can be just nuts-and-bolts issues; however, they are still very important to the successful operation of our systems.

If you are a chief officer, your job is to identify the important things that your organization stands for and is striving toward. Those important things are stakeholder identification, vision, purpose, and values, and they form the platform on which most organizations build a strategic plan.

As an example, if the objective of employee empowerment is to create a high performance organization in which employees make consistent decisions and behave in alignment with departmental values, it is necessary that employees first understand the organization's relationships with its customers—beginning with identifying who the customers are, why they are customers, what we do for them, and what we perceive they want from us.

Customers are identified as stakeholders because in one form or another they are influenced by or have a stake in our operational decisions regarding service delivery. They are key to our system, and understanding their relationship to service delivery is essential if you expect employees to consider them in decision-making.

The point to all of this is that the leader cannot expect consistent employee decision-making when decisions are being made independently of consideration for the customers who could be affected. Consideration of your system customers is the priority of your organizational decision filter. Identifying the customer and answering their questions is just the beginning; validating your perceptions is the next most critical step and requires total commitment from the chief and staff to plan for listening to community dialogue and taking action on what is said. Comparing the data collected with your original perceptions and incorporating their suggestions, feedback, and criticisms into your planning efforts takes courage.

As the leader of an organization, you are often furthest away from the service delivery yet essentially the most responsible for its successful outcome. This paradox of leadership causes some leaders to hold tight to power and control. However, power and control are most effective when they are given away to employees by leadership.

Getting all individuals in an organization committed to achieving the mission and vision results from empowerment. Employees need to feel they have control of the outcome of their delivery system to have buy-in for successful results. Rules and regulations, while essential to the successful operation of an organization, can hamper enthusiasm and initiative when overcontrolling and should almost never be instituted to address a single situation, as the message to all employees can become that in the absence of the rule they would not behave correctly.

> The leader cannot expect consistent employee decision-making when decisions are being made independently of consideration for the customers who could potentially be affected.

Consistent decision-making and alignment of purpose ultimately are a product of all stakeholders understanding the mission and what behaviors are acceptable to achieve its purpose.

Creating System Ownership

*System ownership is not possible until everyone
in the organization understands their shared
responsibility in a successful outcome.*

Key Points

- System ownership comes from an understanding of shared responsibility.
- Employees need to appreciate how they collectively have a role in system ownership.
- Empowerment creates system ownership by transferring responsibility for service to the people who are closest to the delivery of that service.
- System ownership must start with clearly articulated position responsibilities.

If you have ever studied the Marine Corps training philosophy, it becomes apparent very quickly that it works hard on culture. In fact, many of the concepts focused on in this book are consistent with a model used by the Marine Corps.

In the Marine Corps, a recruit is assigned to basic boot camp and is immediately taught the Corps' values so that the basis of decision-making is consistent among all Marines. In addition, everyone knows what the mission is and how they are to be involved in helping the organization accomplish that mission.

From a training perspective, every platoon member is taught how to first be an infantryman and why that job is important to the mission. If any person in the platoon goes down, someone else can, and is expected to, step up and do the work. In addition, frontline officers are empowered to make decisions regarding platoon operations. However, the plan can fail if the leader does not lead employees to value system ownership from a macro perspective instead of just from their much smaller field of responsibility.

I recently discussed system ownership with my battalion chiefs and mid-managers. The discussion was about the result of using overtime for staffing apparatus (something I am sure all of us can relate to these days). The operations chief had directed employees to, if possible, keep apparatus fully staffed. Several of the battalion chiefs had days when they called back and paid overtime to a very high number of employees because of sick and scheduled leave. In one case, we had nine people on overtime in a system that can only pay for about two per day during the year. The bottom line was that we had spent 50% of our yearly allotment of overtime in the first three months of the year.

As we analyzed what occurred, the senior staff assumed responsibility for providing inadequate direction about staffing. The battalion chiefs and the other mid-managers in this case were doing what they were told, to keep the equipment in service if possible and staffed accordingly. We, as leaders, failed to identify the boundaries within which this direction could be applied.

> It is no small task to get everyone on the team to understand how their direct involvement helps the team move forward, but it is essential in an empowered system.

The meeting focused on system ownership and the way the battalion chiefs viewed overtime as something for which they were not responsible. The attitude was almost like, "Well if the money runs out, someone will get some more for us."

I wanted the battalion chiefs to recognize that they were responsible for the use of overtime within the system. I wanted them to take ownership of the system and understand that although it is our department's overtime, it is their responsibility to manage it properly.

This value in system ownership extends to other aspects of the organization and contributes to its overall maturity. In our line of work, we tend to worry only about the progress of the people who report directly to us. If this attitude is allowed to become part of the overriding organizational culture, people will lose focus of the overall team effort. Success, then, becomes something the supervisor views as important only for that functional area.

It is no small task to get everyone on the team to understand how their direct involvement helps the team move forward, but it is essential in an empowered system. Like many leaders, I believe that empowered employees will do the right thing in situations in which they have been given adequate guidance, direction, and training.

It is essential to help employees understand that leadership by empowerment transfers much of the system's operation to the folks closest to providing the service. This also correlates with a tremendous responsibility for employees at all levels of the organization, because to be successful we all must pull our weight. It demands nearly continuous engagement in system operation by all personnel. In many cases, success will depend on having members who are not afraid to hold others responsible when they are not living up to their end of the bargain.

Creating system ownership is the result of the leader's effort to clearly articulate position responsibilities through individual position performance objectives that align expected employee behavior with the organization's leadership philosophy, mission, values, and goal.

Some leaders may balk at pushing too much responsibility downward in the organization. These types of leaders are afraid to empower their employees for fear they will be held accountable for something they view as out of their control. The fact is that, as the leader, you are going to be held accountable regardless. Look at yourself as a servant leader, and your employees will see you as part of the team. Perhaps they will even do the right thing when operating where there is no established policy.

Empowering leadership is the platform from which system ownership is created. It is the direct result of an understanding of shared responsibility. In effect, all employees need to appreciate and understand how they collectively have a role in the organization's success.

Creating system ownership is the result of providing employees with strategic direction so they will know what is important to the organization and how to best apply their efforts to help the organization be successful. Providing employees with organizational values helps define boundaries of behavior and consistent decision-making. Add to this system a full understanding of time allocation (see chapter 16) and you as a leader will be well on your way to creating an empowered organization through system ownership.

15

Value-Based Decision-Making

*Once strategic thinking forms the basis of your organization's
forward progress, the alignment of decision-making and
consistent behavior evolves out of a set of shared values.*

Key Points

- Shared values create consistent behavior.
- In the absence of organizational values, people will use their personal values to make decisions.
- Values must be a part of a decision filter that also includes the organization's mission and vision.
- Shared values help explain decisions in the absence of policy.
- Inconsistent outcomes in similar situations occur when personal values are used in organizational situations.

Up to this point, we have discussed many methods to align employee efforts. The real trick in alignment of employee effort is to make sure decisions that drive organizational efforts are implemented based on consistent values being used in the decision-making process.

The reason organizational values are critical is that if you do not have them, people will use their own personal values as a filter to

> The reason organizational values are critical is that if you do not have them, people will use their own personal values as a filter to determine right from wrong or to make choices when new opportunities present themselves.

determine right from wrong and to make choices when new opportunities present themselves.

Review a case of inconsistent decision-making and I will show you how someone's personal values were used in making decisions regarding the organization. It is not that their personal values are inappropriate; it is that they might not be in alignment with the department's stated values.

In one of my departments, we espouse honesty, integrity, compassion, and trust as our organizational values. These may happen to align with your personal values; they are what employees at this department reported to value most in their leaders and themselves. Our employees want leaders who are honest with them, will support them, are compassionate when mistakes are made, and above all, have integrity in their personal and professional lives. In turn, our leaders expect these values to be demonstrated by employees.

Do not minimize the importance of values. When you are expecting *consistent* decision-making and *consistent* behavior from employees, you must discuss what these words mean and how they are used to guide employee behavior and decision-making. As the leader, you must guide all employees in the organization to use these values in their decision-making processes.

Leaders' inconsistencies occur most frequently when personal values are allowed to prevail in work situations. Imagine how powerful a tool you will have given your employees when they consciously begin to consider the departmental values when making choices. Add the results of your strategic planning effort, as discussed previously, and your employees will have a decision filter that takes into consideration your organization's customer groups, a vision that follows what you are trying to achieve, and a purpose that describes the specific outcomes of your work efforts. These are powerful tools to help create consistent decision-making at all levels of the organization.

Merriam-Webster defines *family values* as "values, especially of a traditional or conservative kind, which are held to promote the sound functioning of the family and to strengthen the fabric of society." Individually, values are reflected not just in what we do but also in who we are in the first place. They represent the very foundation of basic self-belief and constitute the filter we use to make decisions and determine the appropriate resulting behavior. Although hidden from the plain view of others, values and how we apply them are the cornerstone of effective leadership.

In the book *Results-Based Leadership*, the author Dave Ulrich states, "Lacking clear values, rudderless leaders constantly shift from goal to goal. With values, while actions may change, the overall direction and focus stays clear. Values form desired results when leaders ask the question, what is the right thing to do based on this set of circumstances?"

The leader who is responsible for the organization must set direction by identifying and articulating the values that will carry the organization forward. This is where consistency becomes important because if the leader's values shift from decision to decision, the organization will spend valuable time trying to determine which value is being used by the leader, what is important to the leader, and more importantly how and when to follow.

Members of our organizations are expected to make decisions daily that affect the way we are perceived by both our internal and external customers. The values established help employees talk, behave, and deal with these customers. As the leader, you have every right to demand that employees within the organization interact with each other in a manner that speaks to these values. In addition, you should welcome that you will be held accountable for them as well.

Decisions that are sound and based on the same values will align with each other, even if they are made by different members of the organization and even if circumstances vary. This alignment helps address the one thing that will cause us to quickly fail: inconsistency. Inconsistency in the way we act and behave is a clear signal to members of any organization that the leadership is rudderless.

Starting from the premise that we want the same things from each other regardless of our position in the organization really levels the playing field. There is no hierarchy in honesty, as it is expected at every level of the organization from the CEO right down to the janitor who empties the trash cans.

Where we seem to go wrong is at the point where we believe that our organizational values are for everyone else but us. Or worst yet, we use our own personal values to make organizational decisions. Regardless of the situation, personal leadership and the relationships that develop because of it are a key element in the soil that forms the foundation for our plants and trees (read "employees").

One cannot underestimate how critical organizational values are to success. The greatest example can be illustrated by following the complexity of a problem through the organizational chain of command. The process goes something like this: The employee goes to the supervisor and very politely asks a question regarding a policy that they believe is not being applied correctly. The supervisor listens to the issue and then compares what the employee says to how existing policy is written. If no policy exists to directly answer the employee, it is referred up the chain to a higher level. The bottom line is that when push comes to shove, first-line supervisors answer yes and no and right and wrong questions. When gray exists, they instinctively look to folks with more experience.

At the next level, perhaps someone in middle management, the circumstances of the situation are revisited. The middle management employee will have the benefit of experience in their favor and therefore may well have dealt with this type of situation before. If so, the decision is made, and all is communicated back to the first-line employee.

But what if there remains some gray area? What if you are in uncharted territory? What happens when there truly is no right or wrong answer, but rather one in which you are dealing with deciding the best out of many poor options?

As only a person who has been there can attest, when these situations make it to the top, there are no right or wrong answers left. There is no black and white but rather disgustingly depressing various shades of gray. No easy answers are left when issues make it to the top of the organization.

Early in my tenure as a fire chief, I was presented with a problem concerning our new promotional process. The organization had worked hard to develop a process that tried to identify the kinds of behaviors that would be required at the levels we were testing as opposed to grading people for how well they did at the job they were leaving. This situation plays out every day across our country, as most of our promotional systems are a detailed examination and assessment of the position someone already has. We had worked hard to change that type of system and in fact go to a system based in part on a personal accountability assessment, which is a long way of saying résumé.

> There are in fact no easy answers when issues needing to be resolved make it to the top of the organization.

It turned out that one employee had submitted his promotional package and the fire department human resource officer missed the fact that he was short by a few months and did not meet the qualifying years of service requirement. Since this candidate was now number one on the promotional list, it was not a fact very many of the other folks missed.

When it all was laid out, it seemed to most of the staff that we had a cut-and-dried case. Policy says this many years and the employee does not have the years. "This is a slam dunk," I was told, "Just tell him sorry and we will see you next year." "But wait," I challenged the group, "How do you reconcile that this person finished number one? Does that mean anything?" Another officer wondered if we should accept responsibility for the HR guy not catching the thing in the first place. Was the entire process invalid because someone participated who was not eligible? If we disqualified the ineligible person, he would surely appeal to the grievance panel.

With all these questions in mind, I met with the employee, who remained adamant that he wasn't aware of the difference in time between what he had when he participated in the process and the stated requirement. Since honesty is one of our department's core values, and I had no other reason not to believe this employee, I really did believe what he was saying.

I am assuming by now you are starting to fully appreciate how some issues cannot be decided purely by policy or without full consideration for extenuating circumstances. That is the gray area where there are no right and wrong answers, and sometimes the choices you have are ones that could best be described as the least of the bad that is left. This is why values are so important to leading successful organizations.

The tough problems leaders must solve every day do not have straightforward answers, and making any decision may in fact inflame as many people in the organization as it does not. Using a defined set of values helps you conclude and explain how it was you arrived there in the first place.

The leader of an organization needs to move the organization from a regulatory environment to a facilitative one. In other words, you do not want your employees acting entirely because you have rules or boundaries in place. The much more successful approach is to create an environment where employees can act independently and, in effect, align their actions with the desired state of the organization. Using this approach, the leader sets up a facilitative and participative environment.

Over the long-haul, high performance organizations are created when organizational values are used as hiring criteria. In this way, you start out hiring people who share the organization's values, and over time consistent decisions are just a byproduct of everyday organizational business. That is the point when the leader can work on their golf game.

A huge challenge for any leader is to figure out a way to achieve consistent decisions and uniform discipline throughout the organization's many levels, in effect striving to create a situation where the very lowest level of supervision in the organi-

> Having values, using the values to make decisions, and then being able to explain their decisions based on these values is a level of accountability every leader needs to demand from all levels of their organization.

zation makes the same or similar decision as someone at the top when they are presented with a similar set of circumstances. Consistency, in this respect, evolves out of a shared set of values. Having values, using the values to make decisions, and then being able to explain their decisions based on these values is a level of accountability every leader needs to demand from all levels of their organization. Most importantly, that accountability also applies to the leader.

Value-based decision-making helps move your organization from being regulatory and rules driven to holding shared values that create consistent behavior. It may very well be that slightly different decisions are made in various circumstances; however, the basis for those decisions can be articulated and does not violate the main principles by which the organization is grounded. This variance should be appreciated and supported by leadership, as the consequences outside of this approach—that is, always giving the same answer regardless of circumstances—will not produce good outcomes.

Values must be a part of any decision filter expectation, as should the organization's mission and purpose. Using this filter, and a consistent set of organizational values, helps everyone in the organization understand the basis for a specific outcome. Inconsistent outcomes almost always come from situations in which personal values are used to make organizational decisions.

16

Time Allocation

The leader who focuses too much attention on the floor being dirty may never realize that the building doesn't have a roof.

Key Points

- Time allocation is a principle of leadership and management.
- The operating segment is frontline, the improving segment is mid-management, and the creating-the-future segment is senior leadership.
- All employees spend time in each segment, but the percentage of time varies greatly based on organizational responsibility.
- It is easy to retreat to the technical nature of our job when faced with acting strategically.
- Trust should be extended to all employees to operate in all segments.

A system of empowerment results from people in various job functions within the organization understanding their role(s) and what the expectations of others are regarding work outcomes and employee/customer interactions. When these expectations have been clearly stated, accountability and responsibility are natural outcomes.

Fundamentally, time allocation refers to how employees spend their time. Consider that we, as employees, are primarily responsible for operating the system, improving the system, or creating the future. All of us, regardless of rank, spend time in one of these areas. In many cases, we are involved in all three areas at one time or another while dealing with multiple issues.

The time allocation philosophy has nothing to do with gathering information and everything to do with who acts on that information. For instance, use of this model should not preclude the chief or any other officer from moving around the organization to gather information.

Time allocation, a concept that was introduced to me early in my Virginia Beach Fire Department management training, is a visual reminder that in addition to the responsibility of operating, improving, and creating systems, there are considerations for what percentage of time we should spend in these areas based on our position within the organization. The biggest mistake we make as leaders has more to do with the percentage of time we spend in each area than the area within which we are operating at any one moment in time (see table 16–1).

Let us look at a few examples. The largest component of any fire department consists of the firefighters and their frontline supervisors who provide direct service to the public. These employees are operating the system; that is where they will spend most of their time. Leaders can make their jobs easier by giving them tools, empowering them, and trusting that they will do what is necessary to get the job done.

> The time allocation model is a visual reminder that in addition to the responsibility of operating, improving, and creating systems, there are considerations for what percentage of time we should spend in these areas based on our position within the organization.

This is not to say that we do not want our frontline service providers making suggestions concerning improvements or recommending strategic direction, but role clarification specifies the percentage of time they spend in each category of the time allocation model. Certainly, it is critical to maintain a system that is open-ended and in which all persons are engaged.

TABLE 16–1. Time allocation model summary

Behavior	Operating the system (% time)	Improving the system (% time)	Creating the future (% time)
Frontline supervisory personnel	75%	20%	5%
Middle management	20%	60%	20%
Senior leadership	10%	30%	60%

Our middle managers concentrated efforts can be directed at improving current operations. These employees are seasoned and least removed from the frontline service. They are required to coordinate the folks operating the system, and they should be able to make recommendations regarding efficiency and effectiveness of operations.

Chiefs and other senior staff personnel should plan for and create outcomes that will help the organization adapt to the future. This place in the time allocation model is called "creating the future." It is our responsibility to determine the course and provide adequate assets for our frontline employees to operate the system and our mid-mangers to adjust and improve system operations.

It is not the frontline folks who have issues with time allocation or who generally cause system problems, but the mid-managers and upper-level leaders who spend most of their time operating the system. Our employees call this micromanaging. I call it working in the wrong portion of the time allocation model, and it is responsible for many practices that poison our ability to lead organizations.

Think about how we chief officers frequently articulate that we trust our employees to operate the system. Consider the chief who is obsessed with how many hydrants crews painted or why the trucks are not washed. In these situations, you should recognize and respect that the accountability and responsibility for these things fall at some other level in the organization.

I should caution you that it is easy to send mixed messages regarding time allocation, so make sure your employees understand the model is the key to its successful implementation. If you elect to lead by empowerment, you must be willing to let go of some trivial things on the one hand but also understand that accountability for more serious issues remains your immediate responsibility.

Another important aspect of time allocation is that employees at all organizational levels must understand that at various times we are all supposed to manage within every area of the time allocation model.

The truth is the chief may indeed be operating the system; however, if the chief has information that other people in the system do not, the chief has a responsibility and duty to act. If the situation is not critical, one would expect the chief to handle it by finding out why those who are expected to operate the system failed to do so. We must be careful to truly question where we spend most of our time and if being there conflicts with what someone else in the organization is supposed to be empowered to accomplish.

It is imperative that employees be trusted to operate in all areas. The really advantageous aspect of this model is that some of your most strategic thinkers may, at the time, be at the bottom of the organization. You want those members to be able to help the organization by thinking strategically about events that may be a future reality for the department. Remember, we are all in this together, and the difference in being a success or failure may be in how you use all your employees' best traits and skills.

The concept of time allocation will be determined by many factors. If mid-managers are not effective at improving the system because they don't have the required skill set, the leader may have to be more involved in that aspect of the model. In this instance, you may have to elicit the help of some system operators to bring the mid-manager along. The other and more practical thing for the chief officer to do is to mentor and coach the mid-manager into thinking more in line with improving the system as opposed to operating it.

At this point, it should be clear that one of the reasons we are comfortable with doing someone else's job is that if the promotional system works effectively, we have been promoted from a job we have mastered. It is more comfortable for us to do what we have already mastered instead of being challenged to operate in an unknown area.

> At this point, it should be clear that one of the reasons we are comfortable with doing someone else's job is that if the promotional system works effectively, we have been promoted from a job we have mastered. It is very comfortable for us to do what we have already mastered instead of being challenged to operate in an unknown area.

This system is not perfect, nor could one expect we will always be where we are supposed to be in the time allocation model. We are, after all, human. There will inevitably be those situations when you catch yourself doing someone else's job because it may be easier to do it than hold yourself accountable for it not getting done. Keep this last point in mind: we all have a primary responsibility for some aspect of the system's operation and that is where we belong when doing our work.

Tables 16–2 through 16–4 show tasks divided by operational level that are generally associated with operating the system, improving the system, and creating the future.

Time allocation takes leadership and management and divides its functions based on operating the system, improving the system, and creating the future. In general, frontline employees operate the system, mid-managers improve the system, and senior leadership creates the future.

While all leaders by virtue of organizational responsibility should spend time in all areas, the amount of time they spend in each area should vary greatly depending on job responsibility. Mid-managers and upper leadership are the most problematic when considering time allocation because when confronted with improving the present or creating the future it is easy to fall back on the nuts-and-bolts technical nature of our job. Employees at all levels of the organization should be trusted to operate in all segments of the time allocation model.

TABLE 16–2. Time allocation model for frontline personnel

Behavior	Operating the system	Improving the system	Creating the future
% time	75%	20%	5%
Tasks	• Responding to calls • Record keeping • Fire/EMS training • Staffing units • Public education/relations • Employee evaluations • Interactions with other city departments • Preplanning of target hazards • Physical fitness training • Special team activities • Blood pressure checks for the public • Conflict resolution	• Coaching/mentoring • Data analysis • Planning/research • Utilization of Completed Staff Work • Regional cooperation • Resource development • Committee activities • Employee competencies • Service/skill updates and upgrades • Strategic planning • Pursuit of additional education (formal and informal) • Research and development of equipment, strategies, and tactics	• Organizing of daily activities/personnel management • Providing team leadership • Budgeting • Daily equipment/station maintenance • Safety • Special projects • Strategic planning activities • Brainstorming new ideas • Cultivating a global viewpoint • Applying new processes • Determining future needs • Networking • Gathering input from stakeholders • Officer development • Becoming a lifelong learner

TABLE 16–3. Time allocation model for middle management

Behavior	Operating the system	Improving the system	Creating the future
% time	20%	60%	20%
Tasks	• Short-term resource management, including staffing • Responding to calls/field operations • Responding to both internal and external customer needs • Providing a communications link between field personnel and administration • Task-level duties daily • Training	• Coaching/ mentoring • Operational meetings • Critiquing of response performance • Upgrading equipment and resources • Managing strategic plan focus areas/ business groups • Pre-incident preparation of resources • Developing partnerships • Special committees, assignments, and projects • Attending seminars and trainings • Networking	• Becoming a lifelong learner • Succession planning • Alignment with city's mission and values • Visioning and global outlook

TABLE 16–4. Time allocation model for senior leadership

Behavior	Operating the system	Improving the system	Creating the future
% time	10%	30%	60%
Tasks	• Command/ incident management system functions during large/ complex emergency incidents • Addressing immediate safety concerns • Budget development and administration • Training participation • Personnel management/ interactions	• Implementation of policies and procedures • Obtaining resources/ review of budget requests • Interactions with other departments and city administration • Exploring new technologies • Program development • Risk taking • Determining long-term organizational needs	• Strategic planning • Staff development • Secession planning • Mentoring • Career path development • Visioning and global outlook

17

Assessing Organizational Performance

The benefit of using substantiated data from relevant
stakeholders to improve operational processes
is that it specifically addresses the root causes
of issues and not the perceived problems.

Key Points

- Apply the concept of "gold plating" to the fire service.
- Engage relevant stakeholders in assessing organizational performance.
- Design quality/execution graphs.
- Use program quality and execution to gather empirical data.
- Use weighted criteria to prioritize implementation of feedback.
- Use gathered data for other purposes, such as succession planning, job descriptions, and task books.

Early in my career I had an opportunity to attend a program called the Senior Executive Institute at the University of Virginia Darden School of Leadership. During one of the presentations the instructor was explaining what he termed "gold plating." His example was based on project management "mission creep," which is the practice of adding scope to a project that is outside what was originally intended. During the presentation he used the term in a context that was different than project management and more aligned with innovation and cost.

In the instructor's example, he spoke of a video player and talked about the economics of cost versus innovation and how businesses determine their marketing and customers based on how much a product costs, how much innovation the product contains, and how many customers are willing to pay for that level of innovation.

The premise is that marketing professionals know how much a majority of their customers are willing to pay for their products and how much innovation is appropriate for that level of customer. In other words, they understand prior to developing a product how much innovation each available customer is willing to tolerate when it comes to the checkout line. In this process, the cost of the product can be matched to the greatest numbers of customers for that price range.

Products that end up on the high end of the line are referred to as having gold plating, meaning that they are made with a high number of specialized functions that will only appeal to a small number of customers. These products are high end, do a lot of things no average person needs, and are very expensive (fig. 17–1). Keep in mind that innovation is costly and requires a great deal of front-end research and devel-

> Marketing professionals know how much a majority of their customers are willing to pay for their products and how much innovation is appropriate for that level of customer.

opment, and all that R&D needs to be paid back with a profit during the lifecycle of the product.

My instructor at the Senior Executive Institute graphed innovation on a vertical axis along with the decreasing number of customers available for that level of product. The horizontal axis was labeled costs. What we could clearly see was that the higher you went on the innovation chart from bottom to top the more costly the product was and the fewer people the product would appeal to as the buying public.

At the end of this demonstration, the instructor explained gold plating as it relates to the services we provide as local government employees. Because we are taxpayer funded, we seldom think of ourselves as a business that must recoup the cost associated with service delivery. In fact, most of the time we are trying to create more services and, in some cases, do not consider how many people will use those services or whether there is a buying public out there willing to spend their money for those services.

This hit me like a ton of bricks. I had never thought of the services we provide from an innovation versus cost perspective. While I gave all this information its due consideration, in the end, we save babies, right? In all seriousness, how does the cost of a life translate when your example is a high-end video player? Still, there was something that fascinated me about that line graph, so I placed it in my "think about this one day" file and went on about my business.

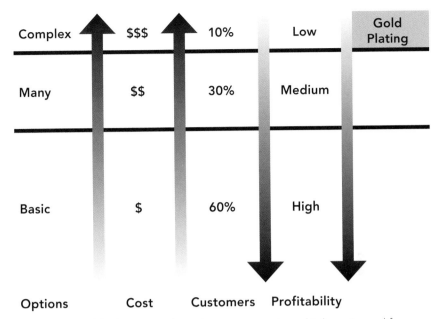

Complex	$$$	10%	Low	Gold Plating
Many	$$	30%	Medium	
Basic	$	60%	High	

| Options | Cost | Customers | Profitability |

FIGURE 17–1. Gold-plating products have many options, high costs, and few customers.

When I was hired as the chief of the Wilmington Fire Department, I was coming off a stint as an assistant county administrator. Prior to starting the job, I surveyed the members of the department as to the culture of the organization and knew what kind of things they thought needed attention. As could be expected, much of what they told me had little to do with system operations and most of the results were culture oriented.

As I settled into the job, I managed to get my head around the cultural stuff, and it was then time to pay some much-needed attention to the service delivery system. As I was pondering another survey instrument my mind wondered back to that innovation versus cost graph. Was it possible to use a similar tool and get results pertaining to the service delivery aspects of our efforts? My main desired outcome was to determine the quality of our services and conversely how well our employees execute these services when providing them to our customers.

I have maintained that as fire service professionals we are always trying to identify and fix the next broken thing. It is just in our nature to try to fix stuff, and from that perspective there is always something broken in our organizations. If you do not believe this, then stop by any fire station in this country and have a cup of coffee. What you will quickly figure out is that these folks are not talking about how great things are but rather what makes no sense and is sorely in need of repair. By the way, I love that about our business—we are fixers!

So how could I capitalize on this fixer mentality and provide our firefighters with a different type of survey instrument? That is when I remembered the innovation versus cost graph and got the idea to change it up so that what we asked from our people could be reported as empirical data. My feeling was that most firefighters would be much more inclined to provide feedback as a dot on a graph than some flowering essay that usually ended up telling me what I already knew: staff is screwed up and communication sucks.

After some feedback from my senior staff, we decided that the two most important things we could evaluate ourselves on was the quality of our products and how well we executed when providing those products as services to the public. And so, the Quality/Execution Charting Instrument was born (fig. 17–2).

In a staff brainstorming session, I asked my chiefs to describe our major service categories. Very quickly we wrote them on the board. Suppression, EMS, mutual aid, organizational development, safety, training, clerical, fire inspections,

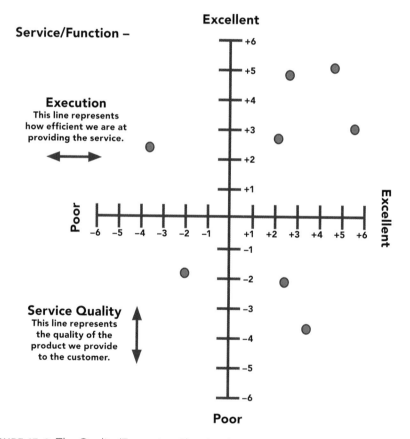

FIGURE 17–2. The Quality/Execution Charting Instrument is designed to illustrate how successful a fire department is in providing services.

data, IT, garage services, budget, human resources, and public education were some of the many service categories listed.

If you are going through this process yourself, at this point you need to do a little more analysis by listing the services in order of their priority to the organization's mission. A useful tool is the weighted criteria decision filter from chapter 21. The weighted criteria will tell you which programs are the most important and additionally you will be able to back up your decisions to whoever wants to know why it was prioritized that way.

Once you have identified and prioritized your programs, place each of them on a sheet of paper with a line graph that has quality on the vertical axis and execution on the horizontal axis. Where the axes cross is labeled as zero, and it counts up from center in all directions. Of course, as the chart shows, the numbers to the left of and under the crossing axes are negative. Using this tool, our folks could place a

> Understanding which programs are of good quality is perhaps the most important feature of this process because who cares if you are executing something well if the quality of the product sucks.

dot on the area of the graph they felt best represented their opinion regarding that service. Further away from the center meant a stronger opinion, positive or negative based on direction.

The four quadrants on the graph are as follows:

- Left Upper: Good Quality/Poor Execution
- Right Upper: Good Quality/Good Execution
- Left Bottom: Poor Quality/Poor Execution
- Right Bottom: Poor Quality/Good Execution

For our purposes, we defined quality as how good we thought the service was, and for execution we used how well we thought it was performed.

Once the results come in you can take the dots on the paper and convert them to numbers, which can be added to a spreadsheet and displayed as a chart. The totality of this information will show you all your service delivery programs and how they compare with each other in both quality and execution.

Don't let it get lost here that understanding which programs are of good quality is perhaps the most important feature of this process because who cares if you are executing something well if the quality of the product sucks. This would be like going to a restaurant and getting really bad food; however, it comes out from the kitchen fast. At the end of the day the food still sucks, and the customer is unhappy.

The other important feature about this type of input is it indicates what is the farthest out of whack. If you have displayed your programs on the chart by virtue of their importance, then you can prioritize which program should be the highest priority for attention. Remember, one of the most important things a leader does is monitor the capacity of the organization to get things accomplished based on the ongoing work and available personnel.

To dig a little deeper, I identified relevant stakeholders for each of these program areas and had them describe for me the individual elements that made up these programs. For example, the suppression captains and firefighters listed all aspects of what is necessary to provide suppression and EMS services in field operations. We did the same for training, fire inspections, and all the other service categories.

The information you gather during both the program and program elements analysis exercise can be used to validate and update your job descriptions. In addition, you can use this information to create a task book that can be used for developing training programs for all ranks in the department. In doing this extra work, not only will you have information about your operational services, but you will be a long way down the road to developing a succession planning and/or career development program.

If you are familiar with fire suppression operations, you already know what kind of information we got back from our frontline folks. In the example, they told us that to accomplish this function properly, we needed to be able to provide size-up, ventilation, forcible entry, hoseline placement, search and rescue, safety, and so on.

Your list may be similar or contain additional elements not shown in this example. The important thing is that the totality of a program can be described as the individual elements that make up that program.

Once all the elements of your programs are determined and prioritized, you can give your quality/execution survey to relevant stakeholders. As the information is returned, it can be analyzed in the same way as was done for your programs survey so that you know, in order of priority, what individual elements of the program need the most work, whether the needed work is in quality, execution, or both.

When you are looking at all this information, do not faint. It looks like a great deal of work and in fact it might be when all is said and done. The choice you have at this point is to either run like hell away from your issues or just decide to peel the onion back one layer at a time. Use your prioritized list of programs and elements, review and validate the weighted criteria to remind yourself why they were prioritized in a specific order, and just start working. If you have a strategic or business plan for your department, this is also a good time to write outcome statements for each of the elements you need to work on and place them in the plan. Doing this will help you keep your eye on the ball when working on objectives that may take a great deal of time and effort to complete.

The beauty of this process is that not only are you acting on empirical input from relevant stakeholders (i.e., working on root cause issues and not perceived problems) but you have included all department stakeholders at some point in the process, and maybe more importantly you can explain why you are working on both the program and the program's individual elements.

Since I first used this type of process, I have found it a very versatile tool that can be used to assess just about any aspect of your organization. If you want to do some real organizational development work, you can even analyze if and why each of the various stakeholders answered based on their computed results. That, though, is gold plating regarding what you expect out of this book, and at the end of the day nobody needs a book that gives them more than they could ever use.

To leverage the fixer mentality of firefighters, you need to develop tools that get to the broken aspects of your organization. The quality/execution graphing exercise allows relevant and knowledgeable stakeholders to provide feedback on your system operations and the individual aspects of those operations that make up its functions.

To recap, quality assessment evaluates how good a program is, while execution asks how well the program is performed. In this case, quality is foremost in importance, as it really doesn't matter how well you execute something that is not very good. Using empirical data to develop strategies to improve your organization will help get to the root cause of service delivery issues. In addition to specifically addressing operational issues, the information can be used to develop job descriptions, training exercises, and task books.

18

Evaluating Employee Performance

We are all ignorant, just about different things.

—Mark Twain

Key Points

- Develop processes that evaluate and respect individual skills and strengths.
- Understand systems that create environments where employees flourish.
- Maximize people's strengths and minimize their weaknesses.
- Typical evaluations can be an assessment of traits based on opinion.
- Typical performance evaluations may cause the employee to chase the evaluator's rating and not success for the organization.
- Clear expectations are important in the evaluation process.
- Create systems that switch ownership of the evaluation process from the supervisor to the employee.
- Use systems that switch from judging to helping.
- Use self-assessment to form the basis of an employee work plan.

We can make a terrible mistake when we compare individuals with each other. As a young company officer, I was approached by the battalion chief, who informed me that he needed to transfer someone from my shift to another shift at a neighboring station. Of course, I did what any young officer would do and went down my list of personnel and very quickly categorized them from what I perceived was best to worse. I then offered up my worst person.

Now mind you, this was not a bad person or a bad firefighter, but when I compared him to all the others on the shift he came up on the short end of the stick. This process repeated itself until I noticed that the person I had originally perceived to be my best guy was now somewhere down the list and in fact not even close to what I considered my best. I also found out that on occasion I really needed the talents of one of the folks I sent away as my so-called worst employee.

Somewhere in the evolution of transfers it occurred to me that my previous best guy was still great but what had changed was how he was being compared with the other members of the shift. This in and of itself proved to be a terrible way to lead, evaluate, and utilize the talent that I had at any one time.

The next time the request came to offer up a person for transfer, I inquired as to the situation that necessitated the transfer. What kind of skills do you need? Is there a specific problem you are trying to address that would require strong leadership? The conversation turned from who would be on the short end of the stick to who is best suited for the position and whose services would best benefit the organization. This way, the organization benefited by getting the right person in the right place.

> As a leader you should have respect for everyone in the organization and work to determine where their talents can best benefit all involved.

As a leader, you should have respect for everyone in the organization and work to determine where their talents can best benefit all involved. Just because some folks have different talents from each other does not mean they are less valuable to the organization or treated with less respect.

> The key is to create an environment where all employees are valued for what they bring to the table and not to establish unrealistic expectations for those who do not have a particular set of skills.

I have always been fond of Mark Twain's quote "We are all ignorant, just about different things." This quote serves to remind me that I am not the smartest or even best talent in my organization. I do have skills that are different from some, but likewise others have skills that I do not. No one is best or worst; we are just different. We are similar in that we all have faults and weaknesses that need to be minimized and strengths that need to be maximized. The key is to create an environment where all employees are valued for what they bring to the table and not to establish unrealistic expectations for those who do not have a particular set of skills. To do otherwise just sets your folks up for failure.

There is always apprehension around performance evaluation time—that special rite of passage where a supervisor and employee can discuss successes and challenges. Or, from a more cynical perspective, the supervisor sits in judgment of a subordinate employee and tells them something they either already know about themselves or something they surely do not want to hear.

Conventional wisdom holds that if your performance is satisfactory during the past year, you will get whatever amount of raise or merit increase is approved in the budget. Likewise, if your performance is unsatisfactory, you won't be eligible for the increase, something that doesn't happen very often, I might add. In fact, less than 1% of our city's workforce is rated ineligible for an increase based on performance.

Performance reviews are sometimes used in promotional processes to determine the ability of a candidate to perform at the next level. How do you think those reports usually turn out? Let's see, a supervisor needs to comment and report whether someone they work with daily is either fit for promotion or not. Chances are the supervisor understands what's at stake for the employee and gives them higher-than-deserved marks. Or, worse yet, they inflate the scores for the person because, after all, they work for the supervisor and if that person isn't good then it reflects poorly.

That said, there is a lot on the line when it comes to our success and the performance review we are given by our superiors—professionally for the organization and personally for ourselves.

Correctly preparing, documenting, and conducting performance reviews is hard work. Many of us do not have the training or expertise to do evaluations correctly, nor do we understand the true nature of what the evaluation should be doing for the employee and the organization. It is also very uncomfortable for most of us to sit in formal judgment of another person, regardless of the relationship.

Typical Employee Evaluations

Most of the time, evaluations end up being one person's opinion of how another employee, usually a subordinate, performs based on a series of listed traits and behaviors. In situations where there is a disagreement between the observations of the supervisor and the feelings of the employee, the process usually has the unintended consequence of becoming confrontational, cynical, and degrading for the employee. It is also usually just as hard for the supervisor to conduct the meeting as it is frustrating for the employee to sit through it.

Typical evaluations have questions in them that require someone to say to what extent another person either works hard, looks good, has integrity, does not violate policy, cares about customer service, stays fit, or any number of trait-based observations dealing with someone's working behavior. In addition, the employees of high performance organizations use values to guide their decision-making. If employees do not live by these standards, then they should not be working for the organization, and you don't need the evaluation process to do something about that. The performance review meeting is most often used as the opportune time to convey this message.

Once employees start to recognize the connection between the evaluator and the carrot, they start performing for the evaluator. It signals the employee that the supervisor's rating, not the accomplishments used to benefit the organization, is the carrot. The problem here is that you do not want employees to start working for the evaluator's rating, you want them working because it is the right thing to do for the organization and their own personal and professional goals.

What Should Evaluations Be Doing for Us?

So, what is an evaluation supposed to do for us? Shouldn't an evaluation be an attempt to document what an employee is doing to help the organization accomplish customer service and other goals? Shouldn't it encourage the employee to become better at their job by identifying goals and objectives that would result in an increased ability to perform their job functions? And perhaps the biggest question of all: Who should own the responsibility for these actions, the supervisor or the employee?

The key to any performance review process may well be in how much preparation a supervisor puts in prior to meeting with the employee. That preparation should start at the beginning of a firefighter's career. It is very important for all supervisors to understand just how critical it is to have clear expectations outlined at the beginning of someone's career, new job, or promotion.

Spend time aligning the expectations of the employee with the goals and vision of the organization. Once a strategy for alignment between the employee efforts and the direction of the organization is accomplished, operations should run smoothly because all actions (performance and otherwise) advance the

> Spend time aligning the expectations of the employee with the goals and vision of the organization. Once a strategy for alignment between the employee efforts and the direction of the organization is accomplished, operations should run smoothly because all actions (performance and otherwise) advance the strategies and mission of the organization.

strategies and mission of the organization.

The process of alignment in this case is not as difficult as it might sound. Employees and supervisors get their clues from their immediate supervisor and from written documents such as their job description or task analysis. Discussing these elements prior to an evaluation and then reaching an agreement initiates a partnership between the employee and their supervisor.

The Switch from Judging to Helping

Many progressive leaders today understand that an important link to organizational success involves how well they can align their employees with the work that needs to be done. Using that as guidance, then why would we as leaders not try to concentrate our efforts on how well employees align themselves to work rather than on individual behaviors that, when separated, mean nothing?

The evaluation process utilized by officers in high performance organizations attempts to switch ownership for positive results from the supervisor to the employee, recognizing that we all have a responsibility to analyze our own strengths and weaknesses and then determine our own future direction and goals. Likewise, getting to the place we say we need to go is also our responsibility and not that of our supervisor.

> The evaluation process utilized by officers in high performance organizations attempts to switch ownership for positive results from the supervisor to the employee, recognizing that we all have a responsibility to analyze our own strengths and weaknesses and then determine our own future direction and goals.

To accomplish this, we start by asking our employees to evaluate themselves on how well they achieved their stated goals of the previous year: "Please prepare a brief review of one or two pages of how you feel you have done this past year in comparison to your established plan. What is the most important achievement in your area of work?"

Then supervisors are asked to articulate how they are planning to use their leadership abilities to help advance the members of their workgroup or department toward achieving their stated goals. This is a big switch, from a supervisor telling an employee how they should behave to the employee telling the supervisor what they are going to do to help the organization. This aids in solidifying the attitude that our job as supervisors is to take ownership of helping others get where they want to go. It also puts an end to the perceived notion that it is somehow the officer's responsibility to ensure that the employee gets what they want.

Another important aspect of the review is having the employee tell you how they will demonstrate personal responsibility for their own future. This is critical for aspiring employees, as it lets them know up front that it is not the officer's responsibility to see they are promoted or advanced to their desired position. It is the employee's own efforts that will lead them where they want to go, and by having to reduce this to writing, they make the first step toward assuming this responsibility.

Lastly, we should ask each officer to outline key projects and related goals that will advance the work of the department. In this manner the supervisor can evaluate goal alignment and, if necessary, encourage goals that will benefit the department and the employee.

A final key component in this type of evaluation is the evaluator discussing with the employee how they are willing to help them achieve their stated goals. In all cases we should understand that while an employee must take ownership of their career, it is the supervisor's job to create the environment in which they stand the best chance of success.

The Feedback Session

When you do get a chance to sit down with an employee to discuss performance, the meeting should very quickly focus on the accomplishments in the past year and the impact the employee had on advancing goals and strategies of the organization. Then—and this is the most important part of the feedback session—focus on the future and what the employee wants to do to advance the organization and themselves. What brilliant ideas have they been holding back? What unbelievable opportunity do they want to turn into something awesome for the department? What was the worst process they encountered in the past year that they now want to focus on improving?

Work Plan or Evaluation?

Using an employee-driven process results in creating a combination self-evaluation and proposed work plan. When each employee develops their own evaluation and work plan, the responsibility for accomplishment then belongs to the employee and not the supervisor. It just makes sense that the employee is the one most knowledgeable about their efforts and how they can best help the organization.

To further advance this point, I think people know themselves better than others do. I also think people want to be recognized for their achievements, and who knows more about an employee's achievements than the employee? The other positive aspect of this kind of evaluation is that it articulates what the employee wants to do with their career so their supervisor can then help them.

Conclusion

The evaluation process I have described is designed to create ownership on behalf of the employee to help them see themselves as responsible for their own actions, career, and the success of the organization. It also demonstrates respect for the fact that we are all different and can use different traits and behaviors to advance the organization. Finally, it gets away from comparing one person to another as a form of evaluating individual worth.

I have conducted this process on numerous occasions with my staff. I always find it informative and refreshing to read the reviews. I personally do not like judging people's traits and instead would rather help them get where they want to go, using their best strengths to help them get there. The other positive aspect of this process is that now I can help them because they've told me what they are trying to achieve.

I am sure you will agree that most of us have great employees who want to do their best work and be better people each day of their lives. As evaluators, it is then our job to mentor, coach, and guide employees so they can reach their full potential in both their personal and professional lives. So, let us take a step back from judging people and start a process of change. My feeling is that employees and supervisors will both feel better about the evaluation process.

High performance organizations use evaluation systems that respect the individual skills and strengths of all their members. In other words, supervisors are taught to evaluate how an employee is doing with respect to the talents they have, not compared with the talents of others they work with.

The purpose of an evaluation should be to set the tone for a conversation about how an individual might use their skills to help the organization. In effect, this self-assessment switches ownership of the after-evaluation process from the supervisor to the employee. These are the types of systems that create environments where employees flourish.

Typical evaluations, where people are judged based on how they perform when compared to a set standard, or worse yet someone they work with, are demeaning and nonmotivational. They are generally based on the supervisor's opinion of the traits and skills they most admire in an employee and not those traits and skills that are required for the employee to be successful in the job. This problem is compounded by the fact that most supervisors are not trained effectively in how to conduct an employee performance evaluation.

High performance organizations use evaluation systems that encourage the employee to self-assess against stated and agreed-upon criteria that is placed in a work plan. This process can help facilitate a change from the supervisor judging the employee to actively being engaged in how to help them.

Pay for Performance: Chasing the Wrong Rabbit

High performance leaders understand that employees are quick to recognize false manipulation in the form of compensation. Therefore, they instead pay them fairly, and then engage them in open and honest dialogue about what is needed for the organization to succeed.

Key Points

- Leaders should understand how people are motivated to perform.
- Pay for performance systems distribute incentives based on performance success.
- Metrics for determining pay for performance success are complex.
- Pay for performance systems can cause funding issues.
- Forced distribution systems are limited by budgets.
- Information hoarding and self-promotion can be an outcome.
- High performing employees have intrinsic versus extrinsic motivators.

I have only ever been to the dog track one time in my life. I thought it was fascinating to watch the stalls open and the dogs take off like a bat out of hell chasing a mechanical rabbit. The rabbit itself was attached to a long arm that extends over the track and I assume entices the dogs to run faster and faster.

As I watched subsequent races, it occurred to me that these dogs never catch the rabbit, yet somehow each time the stalls open every one of those dogs runs like crazy. I assume from their behavior that perhaps they are not smart enough to figure out that no matter how fast they run the rabbit is slightly out of reach.

Many leaders use a similar process and call it paying employees for performance. For example, do we generally assume that if we get a rabbit (money in this case) out in front of people they will chase it each time the stall opens? In the case of pay for performance programs, these rabbits take the form of pay incentives, rewards, or any number of other enticements we assume motivate people to provide better service to our customers.

The company, business, or organization wants to do the very best it can and assumes that we as employees will do whatever is required to make this happen if they reward us in some form or fashion. The newest of these incentive programs is called pay for performance. You perform and they pay you for it.

If you are considering this type of reward for performance system, you should first ask two questions: What is the desired outcome for the performance pay plan written in behavioral and measurable terms? What data do you need to support that the results are helping you accomplish the identified desired outcome? Paying for performance helps you provide quality customer service, but you need another metric to define what pay for performance success looks like.

Let us look at a couple of real-life issues involving pay for performance methodology. In most systems, the bar for high performance is not set until after scores are gathered because there is normally only a set amount of money to be distributed. If the bar for high performance was established before the ratings were done, then the organization would have no control over costs. Therefore, this system should more appropriately be called pay by bell curve or forced distribution.

Because our folks are not dogs, they begin to understand the math and the limited pool used to pay high performing employees. The net effect is that employees are placed in direct competition with each other to be on the money side of the bell curve. Members are then no longer concerned with teamwork and team accomplishments but rather with competing against their fellow workers. This may cause team members to withhold vital information from each other that may help another team member accomplish a goal.

Keep in mind that when faced with a situation where you or another person could get more money, you will usually choose yourself. Human nature would dictate that you are in direct competition with your peers for favorable rating from the boss. If they look more successful than you, they will get the money. This can very easily create a system full of employees who practice information hoarding and self-promotion.

In pay for performance systems, it is extremely difficult to objectively assess the difference between low-level entry job performance and highly complex executive job performance. For instance, does the janitor who sweeps the floor have just a good a chance at receiving high performance pay as the mid-management project executive?

Most organizations that boast about pay for performance philosophy talk about the positive attributes of paying high performers but give almost no consideration for the negative impacts that it has on the other employees. They may well have created a positive outcome for 10% of their folks (by virtue of giving them more money), while at the same time devaluing the remaining 90% of the workforce. It is troubling at best to consider what can happen when 90% of your workforce is unmotivated to perform at a high level.

Pay for performance systems also do not have a good answer for how supervisors are supposed to determine fairly if an individual's low performance is the result of their inattention to matters or rather a lack of resources for the employee to do their job. Think of a person who has a goal they want to accomplish but some of the elements of success are held by others and outside of their control.

Money is not what motivates people to perform at high levels. The high performing organization should address leadership by focusing on the hard work of developing leaders who can create environments in which people are intrinsically satisfied with work. High performing employees are almost always motivated by intrinsic and not extrinsic factors.

> If the plan is for the company to give leaders a money tool to motivate people, then they should think hard about the notion that money is almost never a motivator in nonproduction jobs.

If the plan is for the company to give leaders a monetary tool to motivate people, then they should remember that money is almost never a motivator in nonproduction jobs. Leaders who use extrinsic motivators, like pay for performance, assume they can manipulate behavior and that motivation is something that is done to others rather than created from within.

Pay people fairly and equally by percentage of salary based on who is on the organizational team and helping the organization achieve its goals. Work hard to recognize effort as it is compared to ability to reward not only team members who have great skills and are able to prove them but also the people of lesser skills who are working just as hard. Create a sense of teamwork by recognizing all the people on the team for the team's efforts. In the end, the reward is to be a part of something special that is created by many different people all moving in the right direction and helping each other get there.

Pay for performance systems are established based on the premise that the organization needs to distinguish between those who are performing at a high level and those who are not. In doing so, these systems create internal competition among persons who are supposed to be equally committed to achieving the organization's success.

Employees are motivated to perform because they intrinsically understand that the mission of the organization is worthwhile. They are almost never motivated extrinsically by money in the form of pay for performance. It is difficult at best to find somewhere in the public sector where successful pay for performance metrics can be evidenced in the form of better service. This may be unique to service industry providers as opposed to production-oriented processes.

The Circle of Problem-Solving Responsibility

Your organization will be most successful when everyone understands their role in recognizing problems and then acts to implement solutions for solving them.

Key Points

- Problem-solving is an organizational responsibility.
- Leadership's responsibility is strategic.
- Completed Staff Work is leadership's key to problem-solving success.
- Employees have a responsibility to participate in problem-solving.
- Employee engagement in being a part of the solution is imperative.
- Management is key to problem-solving implementation and accountability.

There is a responsibility on the part of every employee in an organization to continuously work to solve tough, long-standing organizational problems, as well as routine problems and issues (fig. 20–1).

The key to handling any issue successfully is often rooted in the way employees and the organization interact to problem solve, implement a solution, and communicate the results. This process is a continuous circle of problem-solving responsibility that involves all members of the organization. Within the circle, each level of the organization has an area of responsibility.

First, leadership has a responsibility to search out potential problems and not just react to problems when they occur. Problems will blindside all organizations. In the fire service, we call those "brush fires." Decisions about these brush fire

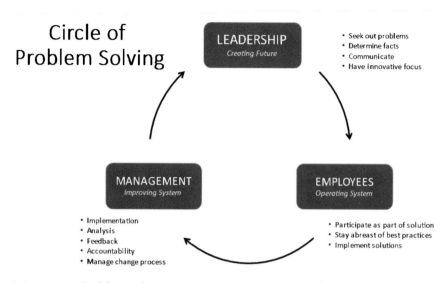

Circle of Problem Solving

LEADERSHIP
Creating Future

- Seek out problems
- Determine facts
- Communicate
- Have innovative focus

MANAGEMENT
Improving System

- Implementation
- Analysis
- Feedback
- Accountability
- Manage change process

EMPLOYEES
Operating System

- Participate as part of solution
- Stay abreast of best practices
- Implement solutions

FIGURE 20-1. Problem-solving in an organization is dependent upon leadership, employees, and management.

issues sometimes must be made in a matter of minutes; however, this type of problem-solving will ultimately overwhelm the organization's ability to be pro-active. Proactive leadership seeks out weak spots and handles them prior to them becoming out-of-control emergencies.

Leadership also has an initial responsibility to determine the facts in a situation and filter out any emotional aspects that tend to cloud the true facts and circumstances. This process is related to analysis of the root cause, which is a part of what you may know as Completed Staff Work (see Chapter 22). Great organizations are committed to the process of Completed Staff Work because it involves all stakeholders, relies on problem-solving through research of facts and data, and provides multiple options as possible solutions for implementation.

Lastly, in this leg of the problem-solving circle, leadership has a responsibility to communicate the implementation of the solution. Since open communication should be a basic value in any high performance organization, the solution in most cases is already widely known prior to its being implemented.

The second leg of the circle of problem-solving responsibility is employee driven. Firefighters are fixers. They are always focused on the next broken thing. These folks are close to the service delivery and while they may not fully see all aspects of a problem or the full consequences of potential solutions, their hearts are almost always in the right place. As an employee you have a responsibility to let leaders know when something needs to be fixed but also to be prepared to offer a potential solution to the problem.

Employees have a responsibility to participate as stakeholders in solving the problem. You simply must see past just your responsibility to recognize problems and, more importantly, see yourself as a part of the solution.

I think it is important to make the distinction here that all members have a responsibility to stay abreast of changes and happenings in their departments. We should all work hard to stay engaged in what is occurring in our organizations. Employees must also be responsible for implementing the solution to the best of their ability. After a full faith effort, employees need to provide the organization with feedback on how the solution to the problem is working.

The circle comes back to management for implementation, analysis, and feedback. Management—in this case, the same folks referred to in the time allocation model as those responsible for improving the system—should evaluate solution results to see how they compare with the solution's desired outcome. In the implementation leg of the circle, management has the responsibility to hold people accountable for the implementation of the solution. Nothing is successful in problem-solving without accountability on the part of all stakeholders to see positive results. As with any process change, to solve problems, management has the responsibility to help all employees manage and understand the change process. This includes how to manage employees' expectations of the results.

> As an employee you also have a responsibility to participate as stakeholders in solving the problem. You simply must see past just your responsibility to recognize problems and, more importantly, see yourself as a part of the solution.

Lastly, management must provide feedback to leadership so modifications can be made if needed. Leadership can then modify the strategic direction should additions or changes in process need to be made.

To recap, everyone has a responsibility for problem-solving regardless of where they are in the organization. Much like the time allocation model, operating the system, improving the system, and strategic direction folks all have specific responsibilities regarding problem-solving. The key is to make certain all people understand their role in the problem-solving circle and are engaged to be a part of the solution.

For organizations to be high performing, all members must see themselves as a part of the problem-solving solution. This process ranges from recognizing current and potential issues, doing the Completed Staff Work to recommend a solution, implementing and providing accountability for implementation compliance, and then providing feedback of how the solution is accomplishing its desired outcome.

Employees, those closest to the service delivery aspect of your business, should be aware of needed changes in process or policy because they are seeing firsthand what is occurring. Suggestions for change should include potential solutions or, at the very least, the willingness to participate in the Completed Staff Work part of the problem-solving analysis.

Management should be focused on implementation, coordination, and account- ability of solutions to problems. They remain a critical connection between the frontline folks and the strategic leadership of the organization.

Leadership, acting as the strategic compass of the organization, must make certain that communication pathways are open and information can flow freely and without interference both up and down the organization.

Strategic Planning

The general who wins a battle makes many calculations in his temple before the battle is fought. The general who loses a battle makes but few calculations beforehand. Thus do many calculations lead to victory, and few calculations to defeat: how much more no calculations at all! It is by attention to this point that I can foresee who is likely to win or lose.

—Sun Tzu

Key Points

- Successful strategic planning connects your customer to your vision.
- A charter is a contract between stakeholders to accomplish the final product.
- The identification of customer groups defines relevant stakeholders.
- Interrelationship diagraphs identify key drivers.
- Vision, purpose, and values form the platform from which alignment occurs.
- Data collection forms the repository for current reality.
- Success indicators form the basis for strategic measurement.
- Weighted criteria prioritize strategies objectively.
- Gap analysis defines the area between current reality and desired outcome.
- Implementation plans are both strategic and objective.
- Adaptative planning is a compromise to exhaustive and thorough study.

If you are the leader and have not developed a strategic direction for your people based on a defined vision and purpose, how can you expect them to help you get there? It would be like starting off in a boat with no clear purpose for your journey or destination. The people in the boat may be onboard, but they will not be able to help you chart a corrected course if you get off track or know why they are there when the journey is over.

Hire a professional facilitator who can help you avoid the pitfalls of the strategic planning process and produce a solid document for your citizens, organization, city manager, and elected body. The real key to successful strategic planning is connecting your customers to your strategic vision. This is accomplished through understanding your relationship to them and their needs. This chapter details a tried-and-true eight-phase strategic planning process for your consideration.

> The real key to successful strategic planning is connecting your customers to your strategic vision.

Phase One: Team Charter

A charter is a written contract that outlines expectations of performance. Every great strategic planning effort starts first with the team that is chartered to produce the product. The *team composition* should include a broad range of people with many different strengths and experiences. In addition, the team should be diverse when it comes to race, gender, and rank. Do not even for a second think you can just get your senior staff in a room and punch one of these plans out. If you use that approach, there will be no buy-in from the rest of your organization.

The charter document includes mention of the rationale for team composition and the specific interrelationships and roles of its members. It is important in this step to decide on delicate issues, for example that people should be free from fear of speaking their mind or that the chief for this process is just another person at the table.

The *team's mission* should also be clearly stated with wording that includes an action verb and the "in order to" format. This method is important to make sure team members understand their deliverables and what the product should achieve. The *team process* being used should also be defined in the charter, as it indicates the steps being used and the format for the final product.

Two additional critical steps in the charter are the *guidelines and boundaries* and the *accountability* sections. These steps are used to make certain the team members are aware of their commitment to the process. The final aspects of the charter include identification of resources/support and then *signature lines* to formalize everyone's commitment. (See an example of a team charter in appendix C.)

Phase Two: Strategy/Interrelationships

For the purposes of this section, I would like for you to think of stakeholders and customers as synonymous with each other. Let us just agree that they are the same thing and move on to the important business of why they are important.

The identification of customer groups spans a broad cross section of the services you provide. For instance, the customers in one situation are the firefighters (internal) who provide a service and in another are the citizens (external) who receive that service. Additional examples include councils, city managers, department heads, religious institutions, higher education providers, social groups, and many other people and groups that have something to do with your organization in any way. Categorize them according to whether they are an internal or external customer.

The next step in the interrelationship section is to have the group provide feedback on why the customers are customers. It is important here to have a clear understanding of what it is you do for them and what it is you want from them. From here you can *affinitize* them into customer groups.

The term "affinitize" can best be described as grouping many customers together by virtue of what service they provide. An example would be the Rotary and Lions Clubs. They are different entities; however, they are both social groups. Therefore, they can be grouped under one customer group called "Social Organizations." In a similar fashion, you also find that institutions that provide instruction can be listed in an "Educational Institutions" customer group. In most cases, when you have completed this process, you will end up with 10 to 20 customer groups.

The next step in this section of the planning is to complete an *interrelationship diagraph*. To do this exercise, place all your customer groups on a piece of paper so that their names form a big circle. Have the group take green and red markers and then start with one customer group and ask who drives who in its relationship with every other customer group. Green arrows can be used for the driver designation, and red indicates they are being driven. Complete this for each customer group as they are related to each other customer group. You will clearly be able to see that certain customer groups are drivers in your relationship while others are being driven. Analyzing the result of this work can be insightful when talking about relationships and considering the potential impact a proposed plan will have on each customer group.

The work is just beginning, as you now need to explore each customer group. This is called *customer analysis*. Examine each customer group and ask the following questions: What do they want from us? What can we do for them? What can they do for us? What do we want them to do for us? Lastly, I always have a category called "cool things to think about." These are ideas noted by team members that may not fit as responses to one of the standard questions but are notes you want to consider at some later stage of the planning process.

If you find your group is really into all of this, now is a great time to use this information to develop a *customer guarantee*: a statement that defines what your customer can expect from you and that can be used as a tagline in your future marketing efforts. An example would be the statement "Committed to earning your trust by always being available during your time of need."

Phase Three: Alignment Phase

The alignment phase, when completed, uses the vision, purpose, and values to create alignment in employee thinking, divisional work, and organizational behavior. Think of it as taking disorganization and conflict and turning it into organization, alignment, and progress.

It is imperative to begin the process of alignment before you start on your plan's implementation. To do anything else would be like handing each of your employees a different playbook for what it is you want to achieve. The net result is everyone doing different things, no prioritization of effort, and no emphasis on what strategies are important.

Now that you know for certain what your customers, including your staff, want from your department, you are ready to work on developing your vision, purpose, and values.

A *vision* is some objective the organization is striving toward. For instance, in Wilmington our vision statement is Excellence Through Service, which is not necessarily a place but something we strive for each day.

The point of a vision statement is that all employees have in front of them a shared dream to guide their actions and behaviors. This is very powerful because it allows the chief and all supervisory personnel to begin the process of aligning employee actions so that they contribute to achieving the vision.

Survey people and ask the question, If the organization could reach its highest potential, what contribution would it make? You can use this information to develop a vision statement. Consider that it should have ideality (desired future), uniqueness (what sets you apart), future orientation (looking forward and proactive in thinking), and imagery (picture or creating of future).

A *purpose statement* professes to all the reason you exist and the actions you will be engaged in to help you strive toward your vision. In the Lynchburg Fire & EMS

> The point of a vision statement is that all employees have in front of them a shared dream to guide their actions and behaviors, one that will ultimately contribute to achieving the vision.

Department our purpose was "To form partnerships that cultivate a safe environment through education, direction, and resolution of fire, emergency medical, or life safety situations." The innovative chief officer knows that these are not just words on a wall. They can and should be used as a tool to align employee efforts.

The best example of this in use is when an employee suggests a new program or community initiative. You should use the vision and purpose statement as a primary and secondary filter to see if the proposal aligns properly. This filtering effort alone ensures your organization works on the right things for the right reasons and helps keep the organization from chasing too many proverbial rabbits we cannot (or will not) ever catch. The bottom line is your people may be doing different jobs, but they are all in the boat and at the very least should be headed toward a common destination.

Now the big one: *values*! The point to department values is that they align employee behavior based a standard set of criteria. See chapters 5 through 7 for a detailed discussion of organizational values and traits.

To recap the alignment phase, phrasing your results should be enough to tell your organization what you are striving to achieve (vision), the types of services you will be providing (purpose), and finally the behavior that will be consistently expected while performing the work (values).

Phase Four: Collection of Data

The essential elements of your efforts in collecting data will be to determine your current environmental reality and assess your organization through the eyes of your previously identified customer groups.

The *environmental scan* of your community is a document that defines your community. What is the physical layout of your community's transportation system? How many road miles, rail line miles, and navigable water miles are there? What kinds of products and how much of these products move along the system during a defined time? What kinds of incidents result from this activity? Another topic might be educational institutions. How many do you have? How many students attend these institutions? Is there a substantial increase or decrease in call volume based on the school's schedule? These are only two examples of what topics can be included in this document, which can be as comprehensive and complex as you are inclined to make it. The point is the final document will be a picture of current reality for your community.

This is an important step in your strategic planning process because the many discussions that will take place over the several months of the planning process will include suggestions based on the perceived notion that one thing or another needs to be addressed. Many times, a document like this can help you quickly distinguish between what is perception and actual reality.

The next exercise is to elicit input from your internal and external stakeholders as to the *characteristics* that make up a great fire department. We all know of the fire and EMS agencies that are on the forefront of our industry, and your goal is to see if their characteristics can be modeled in your department. As an example, you might know that so and so department gets 90% of their scheduled inspections done in a certain time and this has been determined to be world-class service. Visit or contact that department to see what they are doing that you could also implement.

As you did with characteristics, determine the *innovations* that are happening in your industry. Seek out other departments that you feel are on the forefront of modern fire and EMS delivery and see if anything they are doing can be done by your department. Do not be ashamed to steal a good idea, as only one department at a time figures out the best way to be innovative. The rest of us just copy that innovation.

In previous planning efforts, I have used a tool called the Focused Conversation (originated as a military instrument and then reimagined several times by various organizations) for external groups to get feedback on expectations for response time, services, and perceived quality of service, as well as an organizational development survey to examine the culture of the department and the soft side of leadership for internal stakeholders.

The organizational development survey indicated here does not include an assessment of your operations. That part is a separate and independent analysis that includes deployment analysis, response times, reaction times, effective response force times, and apparatus reliability factors, just to name a few.

Phase Five: Success Indicators and Measures

By this point in the process, you will have identified many different strategies as "could be" actions. You also understand the enablers and restrainers that need to be addressed for you to reach your desired state. This is the point where most strategic planning teams, especially fire teams, hit the wall, that proverbial place where everyone loses energy and many of the team members think that they have the fix for the organization's problems, so why beat a dead horse?

To some degree they will be right. You could stop here and have many good things to do over the next several years that will all make your organization better. The problem here is twofold. First, you will not know when you have achieved success at implementing the strategies, and second, you have no way to prioritize what should be worked on first and what could wait a few months. Remember, one of the leader's jobs is to match the organization's work capacity to what needs to be accomplished.

To have appropriate measures in a strategic planning process, the team must be able to articulate a vision for success. In other words, how will the organization know when it has been successful? This is described as *identifying success indicators* and requires the team to develop *measurement tools* that can be used to describe progress. Sounds easy, huh? Now you know why your team might hit the wall.

The next part of the process is called the *weighted criteria decision filter*. This decision filter is developed based on the team's shared expression of how an opportunity may affect your organization, customers, members, image, budget, safety, and resources. In addition, you should consider how the opportunity will align with your mission, whether it will be politically palatable, and to what degree it is a priority. See table 21–1 for a sample weighted criteria chart.

While these elements are all important, they are certainly not all equally so. That is why you should collectively decide what percentage of your department's focus they should occupy. For instance, your customers might make up a 20% value while budget considerations may account for only 10%. The point is that all your prioritization elements need to be collectively weighted to represent 100%.

Additionally, you will need to take each element and give them a scale from 1 to 5 for a rater to assess them on. In this case, 1 would indicate no positive impact and 5 would have an extremely great impact. Obviously, scores between these figures could be calculated accordingly.

As an example, when you are deciding the impact of an opportunity, you could give it a score of 2 for external customer impact which has a 15% weighted

TABLE 21–1. Sample weighted criteria chart

Criteria/ percentage	Impact on mission (20%)	Politics (10%)	Financial impact (10%)	External customer impact (15%)	Alignment with mission and purpose (15%)	Internal customer impact (5%)	Measurability (5%)	Safety (20%)	Total
Defined area	To what degree will this help us achieve our mission?	How will it be perceived by our partner stakeholder groups?	Is there a cost that is not proportional to outcome?	How does it affect our member stakeholders?	How well does it align with our stated mission and purpose?	How does it affect our board of executive directors?	Can the strategy be measured?	To what degree does it help with stakeholder safety, health, or wellness?	
Rating	1 Negative 2 3 Neutral 4 5 Positive	1 Negative 2 3 Neutral 4 5 Positive	1 High cost 2 3 Some cost 4 5 Low cost	1 Negative 2 3 Neutral 4 5 Positive	1 Little 2 3 Some 4 5 Greatly	1 Negative 2 3 Neutral 4 5 Positive	1 Not easily 2 3 Neutral 4 5 Easily	1 Little 2 3 Some 4 5 Greatly	

importance in your criteria worksheet. The formula in the cell would then calculate the score of 2 as the correct percentage of the 15% allocated for that criterion. Likewise, this would be done for each element and for each opportunity.

Collect all opportunities in a spreadsheet and have your team individually rate them. For that matter you could also enlist the help of many different stakeholders in your organization. I would just caution that if you allow five firefighters to participate you allow five members of all the other groups to participate as well so that no group has an advantage when determining the outcome.

When this exercise is complete, combine the individual scores for each opportunity and sort the opportunities from highest to lowest. Your raters have thus solved your dilemma of which opportunities should be considered most important for your organization.

This may seem complicated, but when you get the knack of setting something like this up, its value in helping you identify priority opportunities is immeasurable. It also creates a final product that is an expression of many people's shared values. The collective input gives you good justification when someone asks why you are prioritizing an opportunity because you can state that something ranked high or low from many people rating it across several elements. This alone can help you lay the groundwork for all oral and written communications regarding that opportunity.

The last step in the phase is to take all these opportunities and place them in focus areas. What should become clear is that many of the opportunities are similar in nature. For instance, they might cover organizational development, operations, or business management. Define these overarching groups as *focus areas*. They are the main categories for all your opportunities.

Phase Six: Gap Analysis

Your decision filter work will result in "should be" actions: the highest priority items with the highest percentage scores that should be focused on to meet the new desired state. In other words, these are actions that should be placed in a strategic plan as future direction.

Gap analysis was taught to me early in my career as a process to analytically get from current reality to desired state (fig. 21–1). We used the tool in Virginia Beach after it was brought to us by a consultant named Lee Pitman. This process walks you through identifying a desired state and the performance indicators and measures that will demonstrate success, and then works backward to identify current reality, stakeholders, root cause analysis, and enablers and restrainers.

Gap Analysis Tool

Current Reality	Actions	Desired Outcome
"Begin with the end in mind" —Covey	*Current Reality* → Actions → *Desired Outcome*	
Current Reality	**Actions**	**Desired Outcome**
Brainstorm, identify, and collect data concerning "what is" for this desired outcome.	**Could Be Actions** Brainstorm (dream) the possibilities.	State the Desired Outcome with an action verb and ending with the "to do" statement.
Stakeholders Ask: Who's in the Frame? What do we DO for them and what do we WANT to do for them? What do we THINK they WANT us to do for them? WHY?	**Enablers & Restrainers** Role-play key stakeholders/ customers and ask: What will enable us to achieve this desired outcome and what will restrain us from reaching this desired outcome?	**Performance Indicators** Ask: If we are successful in achieving this desired outcome, how will we know? When...
Root Cause Analysis Separates "perception" from "reality" by asking WHY 5 times.	**Timeline** Who, what, when, where, how?	**Measures** Develop a measurement tool for each indicator and collect data. **Communication** For each action based on Stakeholder needs

FIGURE 21–1. Gap analysis tool

Take the example of losing weight. I might establish that my desired state is to lose (action verb) weight "(in order to)" be healthier and live longer (desired outcome). I can then establish a performance indicator that says I will know I am successful in my efforts if I get to 185 lb. and off my high blood pressure medication. At this point I would also like to communicate to my wife, as a key stakeholder, that my desired outcome will likely affect her meal planning in the future.

From there I go back to state what is current reality. Let's say I get on the scale, and it says I am 215 lb. That means I have 30 lb. to lose to achieve my desired state. Next, I need to identify the stakeholders who have a stake in my efforts. The list might include an exercise trainer and a nutritionist. Finally, I should conduct a root cause analysis. In this case, I would answer the why questions as to how I got 30 lb. overweight.

Now that I have identified a desired outcome, know a metric that will show when I am successful, and have established current reality, the root cause of my issue, and the stakeholders involved, I am ready to decide actions. Perhaps I could get a gym membership, hire a nutritionist, run a mile each day, or buy a book on healthy living. All of these seem possible but may or may not be, so I must evaluate these actions based on restrainers and enablers. Restrainers keep you from your goal and enablers help you reach your goal. For instance, I might not be able to afford a gym membership (restrainer) because I have a new car payment. I may however be able to get a gym membership if I sell my new car and buy a used one (enabler). Lastly, I need to set a timeline for achieving the desired state.

While this is a simple example, I suspect you can pick up the value of the tool's use in helping you reach a certain desired state by analyzing various data points and deciding on actions that will drive you toward success and eliminate those things that are keeping you from getting there.

Phase Seven: Action Plan

At this point in the process, you may want to stick your head in the sand. I can tell you from experience that the many opportunities, and the complexity of those opportunities, appear overwhelming for the leader and the organization. That said, the old saying that you eat an elephant one bite at a time could not be more applicable. Lay out a plan that is considerate of the workforce's capacity and combines the hard goals that will take a while and the easy low-hanging fruit that can be accomplished quickly.

To tie up the working end of this strategic process I recommend using the Completed Staff Work Matrix to accomplish each of your opportunities. This model can be found in chapter 22.

Phase Eight: The Plan(s)

There is no better time in a strategic planning process than the time you will spend writing your strategic document, called a *strategic plan* or *strategic business plan*. This document needs to be high level, and it is not the place to talk about enablers and restrainers but rather to tell the readers what you desire to do and what methods you will take to get there. This is also the place where you articulate how your plan aligns with the other plans that may already be published by your organization.

I will be the first to tell you that many strategic plans get placed on the shelf when completed and never again see the light of day. Imagine, after all that work, just using the process of planning itself as the end goal instead of going after the many great things that could be done if you could convince people to get down to work. That is where the work plan comes into play.

The *work plan* is the document you use to track the progress of your many projects. To build this plan, place all your prioritized opportunities in a spreadsheet. I am fond of a product called Smartsheet because it allows collaboration between stakeholders and allows you to build dashboards to demonstrate and track progress.

This document can be used to assign responsibility for the task and to track progress as a percentage, the task measure, the steps in the process to accomplish the task, and the status as reported by the owner. Many other elements can be included in the plan. What I want you to take away from this point is that the work plan is where the magic happens. It is the key component that organizes opportunities and serves as a platform for accountability to make them happen. See table 21–2 for a sample work plan.

Adaptive Planning

Success may well be defined as recognizing the need for change before the need for change changes again. So, having first laid out the elements of strategic planning, I would be remiss if I did not tell you that many modern high performance leaders use some of the traditional aspects of strategic planning in a condensed fashion. Think about this like you would the difference between how a rookie is taught in recruit class and what the folks teach them when they get on the line.

I have come to call this new "on steroids" approach adaptive planning. Adaptive planning is necessary because of the nature, complexity, and speed of change. If organizations drag their feet with many of the more complex aspects of strategic planning, they face the prospect that the final product will be outdated once it is completed.

Make decisions and plans on the fly and understand they will only be 80% of what you may need. You should be nimble enough in your organization to move ahead with the 80% you get right and work at the remaining 20% as you go.

Listed below are the steps in the adaptive planning process:

1. Write your vision, purpose, and value statements.
2. Determine customer groups.

    - What do we do for them?
    - What do we want to do for them?
    - What do they do for us?

3. Establish microbusiness units (made up of groups of employees from different aspects of the organization that are brought together to represent the business unit).
4. Analyze strengths, weaknesses, opportunities, and threats.
5. Conduct an external customer survey based on customer groups.

TABLE 21–2. Sample work plan

Focus area				
Focus area goal	The organizational desired outcome of a division as the result of programs, activities, and/or processes.			
Focus area goal owner	(By title)			
Strategies	A program, activity, or process with a desired outcome that contributes to the organization's mission.			
Strategy owner	(By title)			
Tasks	Describes measurable steps, processes, and/or resources necessary to achieve a strategy.			
Weighted rating	Numerical rating based on weighted criteria survey.			
Priority ranking	How is task rated as compared to all other tasks? 1–96			
Priority	1: Very high 2: High 3: Significant 4: Low 5: Very low			
Task owner	(By name)			
Task measure	How will we know we are successful?			
Status	What has occurred? What can be expected? What is the current status?			
Process steps	Describes broad actions necessary to achieve task.			
Obstacles/ challenges/notes				
Task completion dates	2023			
	2024			
	2025			

6. Conduct an internal customer survey using Likert scales to analyze the current reality related to trust, controls, motivation, teamwork, leadership, and accountability.
7. Research innovative organizations.
8. Research industry "weak signals."
9. Evaluate what is successful (what falls off the table and what gets reinforced).
10. Create strategies based on all stakeholder input.
11. Develop objectives/projects to accomplish strategies.
12. Determine common themes and define focus areas.
13. Write your strategic plan.

To recap, the high performance leader recognizes that all members of the team must have a clear vision for the organization's desired outcomes. This is accomplished through the publishing of a strategic plan, business plan, or work plan.

There are many elements of a successful strategic plan; however, the measurement of success can be most effectively gauged on how well the plan demonstrates to internal stakeholders the requirements necessary to accomplish the business's desired outcome. Essentially the process connects your customers to your vision.

The strategic planning charter is a tool used to define the boundaries for team cooperation and responsibility. It is ultimately the contractual arrangement between members of the strategic planning team.

No strategic planning can be done before careful consideration of your stakeholders. In this case, stakeholders can be defined as customers of many different types whom the organization provides a service to or who provide a service to the organization. Interrelationship diagraphs are done to identify the relationships between these customer groups and help establish who is driving whom in the relationship.

Vision, purpose, and values are necessary to establish a platform from which alignment occurs. This alignment is in the form of expectations of actions and behavior.

Data collection is the repository from which current reality is established. This information is used to fine-tune strategies based on the difference between perception and reality.

Success indicators are needed to establish a baseline for understanding success. In this process, the organization answers the question "When will we know we are successful?"

A gap analysis separates current reality from desired outcome. In doing so a gap will become apparent, and the strategies derived from the gap will drive you toward success.

Weighted criteria are established and used to prioritize strategies objectively. This process answers the why question regarding your strategies and how they are prioritized.

Written plans can be detailed to advise the organization of planned direction. It is always better to reduce the various strategies into a work plan, which defines steps for strategy completion and assigns success for its implementation.

Conventional strategic planning can be complex and time consuming. Depending on the organization's starting point and what needs to be accomplished, adaptive planning can be a much more efficient and focused tool. The adaptive planning process will not be 100% thorough but will be responsive enough depending on how quickly a plan needs to be produced and implemented in your organization.

Completed Staff Work

*No project, process, or effort is worth its salt
unless it has been first vetted by those who
are responsible for its implementation.*

Key Points

- Leaders cannot be involved in every aspect of organizational decision-making.
- Employees need to learn processes that can produce solutions to problems that are inclusive of stakeholder input and fully vetted.
- Completed Staff Work is a series of steps used to analyze a problem and produce an effective solution.

Successful organizations create their own future by concentrating on and anticipating issues that affect service delivery. This process involves members who seek out organizational problems and then develop innovative ways to solve them.

Nothing makes a chief's job easier than an organization full of people who are constantly seeking out potential problems and fixing them before they become real showstopper issues. On the other hand, nothing frustrates a chief

> Nothing makes a chief's job easier than an organization full of people who are constantly seeking out potential problems and fixing them before they become real showstopper issues.

more than when these same people seek out potential problems but recommend solutions that do not get to the root cause.

Decision-makers in medium to large organizations do not have time for intimate involvement in every issue, and they therefore must rely on others to do research and report the pros and cons of a particular recommendation. This process can be akin to walking through a minefield without a map: each step forward has the potential to end the journey or cripple the traveler. Even in the best scenario, the decision-maker will be using a map of the minefield prepared by someone else, which requires confidence the map was prepared correctly.

Nothing can get a new officer in more trouble than being asked to study and prepare a report on an evolving issue and not doing good research. Perhaps they do not take enough time to get all the data or, worse yet, jump to a premature conclusion about the problem itself. It could be as simple as the decision-maker moving forward on a project and then finding out that the recommendation was grossly underfunded. It may be as serious as recommending a solution that creates another set of problems and, consequently, sends the organization spiraling into a paralyzed state.

Many of our decision-making problems can be traced to several fundamental fallacies. The first is that very few firefighters are taught how to properly conduct research and report conclusions and recommendations. Second, if no established and approved method is used to solve problems, the results can vary and be as inconsistent as the perspectives of the people doing the research.

Completed Staff Work Process

Completed Staff Work, a concept developed and implemented by senior city management in Virginia Beach, Virginia, consists of a series of steps employees follow when researching an issue: identifying the issue and establishing a work process, collecting and analyzing the data, developing and assessing options, making recommendations and drafting the report, obtaining decision-maker approval, presenting to the approving body, implementing the approved decision, and following-up (see table 22–1).

Each of the steps contains several subsets of considerations. When followed in recommended order, they provide an organized and professional manner for developing and presenting the final report. Using a logical method in which one step builds on the other maintains focus on the issue and generates leader confidence that the final recommendation has been fully researched and all matters of relevance have been considered.

TABLE 22–1. Completed Staff Work Matrix

1. Identify issues and establish work process	2. Collect and analyze data	3. Develop and assess options	4. Develop recommendations and draft report
Determine scope of the issue	Research data and information	Determine alternatives	Seek consensus of stakeholders on recommended options
Develop clear statement of the issue	Involve stakeholders	Develop cost options	Select best option
Identify decision-maker(s)	Seek customer input	Consider budget impacts	Resolve conflicts
Clarify needs and expectations of decision-maker(s)		Identify policy impacts	Consider presentation methods
Assign lead responsibility		Consider noncost impacts	Draft report
Set timetable		Consider public relations and marketing aspects	Circulate draft for feedback (broader audience)
Identify internal and external stakeholders			
Identify resources required			

(continues)

TABLE 22–1. Completed Staff Work Matrix (*continued*)

5. Obtain approval of decision-maker	6. Present to City Council (if appropriate)	7. Implement approved decision	8. Follow up
Present to decision-maker	Determine presentation method	Distribute final report and communicate to appropriate staff	Monitor implementation
Alert decision-maker to unresolved conflicts	Position the "product"	Determine implementation strategy	Evaluate implementation
Determine level of Council involvement	Publish/present		
	Obtain Council decision		

Step One: Identify Issue and Establish Work Process

Determine the Scope of the Issue

If there is a good place to do a good job on project research, it is in the very beginning. Just like building a house, every good project starts with a good foundation. In the case of project research and reporting, this foundation is essential to making sure the researchers understand their responsibility and commitment to the process and that they stay focused on working to resolve the real problem.

The first step in your journey is to determine the scope of the issue being addressed. For instance, is the issue a far-reaching and broad organizational issue, or is the problem of narrow focus, perhaps involving only a limited number of people or small workgroups?

This is a very important part of the process. Spend quality time here, and do not be in a hurry! Problems will certainly arise if you think the issue is very narrowly focused but it happens that the outcome affects more stakeholders than you anticipated. Even worse would be the scenario where the implementation of your work causes more problems than originally existed. If nothing else, taking time in the beginning to understand the complexity of the issue will give you a general idea of how difficult you can expect the process to be.

> If nothing else, taking time in the beginning to understand the complexity of the issue will give you a general idea of how difficult you can expect the process to be.

After determining how broad or narrow the issue is, consider the parameters of the issue. For instance, are there geographical barriers that would preclude a timely resolution to the research? Other parameters, such as fiscal considerations, public relations, and deadlines, need to be considered as potential parameters under which your research will take place.

The next step in determining the scope of the issue is to consider what the final product will look like. Is this a policy report or a suggested ordinance change, or will the results just form the basis of a final decision regarding a particular problem? Understanding what is needed should help you understand the true nature of the issue and whether it is strategic or operational in nature. That will tell you volumes about who the potential stakeholders are and the organizational commitment that will be necessary for you to get started.

Develop a Clear Statement of the Issue

Countless times I have seen a committee start out to resolve one issue and end up working on something completely different. In one case, I was part of a group that was supposed to develop a reduction-in-force policy and, instead, recommended which positions needed to be eliminated. The decision-maker in this case wanted a policy that could be applied across the workforce; instead, the group jumped to the results of how the implementation would occur.

It is easy to jump to a perceived solution without developing and analyzing the data needed to support the decision. The danger is that you may occasionally be correct; however, your solution often will address something other than the real root cause of the problem. Although you may feel good about saving time, the results of your work most likely will put the organization further behind.

For this reason, it is imperative that you start your effort with a clear statement of the issue. A statement focused on an expected outcome is easily understood and concise so the work doesn't stray from its objective. An example of an unclear statement would be, "Resolve the fact that we have too many workers and must reduce the workforce." A much clearer statement would be, "Develop a workforce-reduction policy using criteria that aligns with our vision, purpose, and values."

The first statement does not lend itself to a clear understanding of what the group is to accomplish. Does the decision-maker want recommendations concerning reduction in existing programs or entire departments? Does the decision-maker just want recommendations as to what specific positions to eliminate? In the second example, the statement is very clear about the outcome of the work. In essence, the person assigning this work is asking for a policy that uses consistent criteria formed by aligning the policy with the organization's vision, purpose, and values.

> The most advantageous part of developing a clear outcome statement is that it allows the person assigned lead responsibility and the group members to resolve misunderstandings prior to any major work taking place.

The most advantageous part of developing a clear outcome statement is that it allows the person assigned lead responsibility and the group members to resolve misunderstandings prior to any major work taking place. As work progresses and internal and external forces start pulling the group off target, it serves as a reminder of the group's original mission.

Identify Decision-Makers

In all cases, you will need to consider the identity of the ultimate person(s) or group(s) that will decide on your recommendation—if the objective is an ordinance, for example, the city council would be the decision-maker. Otherwise, for a recommendation to implement or change city policy, the city manager most likely would be the decision-maker. Specifying this is essential because there are threads that tie our organizations together, and each thread has a different set of expectations and needs.

Identifying the decision-maker also allows you to first consider the position and feelings of that person or group regarding the issue. Political considerations or timing issues might make it difficult for a political body to approve your recommendation even if it is the best option under existing circumstances. Likewise, if the recommendation will be submitted to your bosses, you should consider the internal and external pressures that may impact them if they approve your final recommendations. In all cases, discuss the project with them early in the process so that all parties involved have a clear understanding of the issue, the statement of the issue, and the desired outcome.

Clarify Needs/Expectations of the Decision-Maker

Building on the previous work, it is now time to gain some specific information regarding the project needs and expectations of the decision-maker. For example, is the statement of the issue in accordance with the decision-maker's expectations? Does it articulate what the group will be trying to accomplish, and are you heading down the right path?

Another important item to discuss with the decision-maker is when the report is expected to be completed. Nothing frustrates a leader more than waiting on a group's work longer than expected. The project leader and the decision-maker thus need to come to some realistic agreement as to how long the process is expected to take.

Finally, take time to discuss how red flag issues (those not anticipated by the project leader or decision-maker prior to the start of the process) will be addressed. A good example would be finding out that the data collected are pointing toward an outcome that was not expected or is drastically different from previous thinking. Other examples could be changes in the fiscal or political landscape. More than anything else, as the project leader, you will want to know how much autonomy you will have to complete the project and how often the leader wants to be briefed on your progress.

Assign a Lead Responsibility

Although we like to think that a very functional team can work without leadership, experience tells us that is not always the case. Even mature and engaged teams need a designated leader, if for no other reason than to serve as a focal point for collecting and disseminating the group's work.

Whatever your experience has been, it is always best to decide up front who will be the group issue leader—the person assigned responsibility for the project and the outcome of the group's work. In addition, this person will make sure that the group completes its objectives and all deadlines are met.

At this point, the issue leader should convey to the group the level of responsibility and accountability the decision-maker has bestowed on them to get the work completed. For example, it may be necessary for the group leader to delegate certain aspects of the project data collection; along with that, the group leader will need to know what authority they have to see that the work is carried out. At the very least, the group leader can clear up any misconceptions concerning the level of delegation the decision-maker is comfortable giving out to get the project completed.

Set the Timetable

Like all successful projects, yours will need a timetable. It should include information about how your work will be scheduled to ultimately meet the deadline established by the decision-maker and also meet any established benchmarks.

One easy way to do this is to write your projected beginning and ending deadlines on a timeline. From there you can take all your project components and plug them in based on how you feel the project will progress. Refer to your earlier discussions with the decision-maker about how broad or narrow the focus of the issue is. That could give you some indication of the time that will be required. After a little practice with the process, you will be able to accurately estimate which projects are the most complex and which will take the most time.

Consider whether the resolution of your issue will depend on another action or issue being resolved. I can remember a time when our department ran three project groups at the same time because multiple issues were tied together by common threads. A process like that takes much coordination and communication among groups. Using a set plan with timetables for completion gives all groups an opportunity to monitor their progress against the work of the other groups.

Identify Internal and External Stakeholders

Project research involves identifying the internal and external stakeholders who may be affected by the issue you are addressing. A look back at your customer groups you identified in chapter 21 will give you a head start in understanding who these key players might be. When considering stakeholders, ask yourself not only who will be impacted by your work but also who can lend insight and expertise to your effort. By identifying the people who will be most affected and those who possess the technical knowledge and expertise needed, you will go a long way toward establishing key stakeholders and potentially those you may want represented on your project team.

Council members, commissions, boards, and even community issue leaders may need to be included and recognized as stakeholders. Many recommendations fail because egotistical political leaders felt they were left out of the process. Only you will be able to decide the level of participation you want these folks to have. However, I will caution you that no politician or city manager wants to be surprised, unless it is their birthday. If you are unsure, consider at least copying them on the group's meeting minutes.

When deciding who should participate in your group, remember the old saying, "Right people in the right place with the right skills at the right time." This is very powerful because, just like every successful team, your team will need a myriad of talents. Remember, you need not only thinkers but also folks who can write, format, create graphs, analyze data, sort materials, take minutes, and (finally) bring refreshments. Considering the various skills needed during the selection process will ensure that you complete all aspects of the work in a professional manner.

You may also want to consider all levels of your organization if multiple levels exist. I have found success in including members from the newest recruit to the oldest veteran. Involving all levels of your organization also helps facilitate the communication process throughout the organization.

Identify the Resources Required

One of the final considerations of the work process includes identifying resources you will need to complete your work, such as keeping minutes of the group's progress. Disseminating this information to organization members helps remove the first barrier to potential change: fear of the unknown. Be very leery if you are working on something and feel the need to keep it a secret. More times than not, something that cannot be released in the meeting minutes shouldn't be brought up in that setting in the first place.

Another resource consideration is technology. Having people with technology skills is one thing; having the software to do what you need is quite another. Considering up front what you plan to accomplish with the data you collect later in the process may save your group valuable time and the negative consequences of losing project momentum.

Determine the Characteristics of the Project

Do not confuse this step with determining the scope of the issue as completed earlier. To decide the characteristics of the project, ask yourself, "When I finish my work, what will the project look like? Will it be a policy report, an administrative directive, or a presentation to the elected body?" It could even be as simple as a memo to inform the organization of a change in methods or guidelines.

Each type of report varies in the level of detail that will be given in the final presentation. That said, under no circumstances should you fail to complete and document all the steps in the process because of the type of report. Remember, Completed Staff Work is not so much about what the final report looks like as it is about the complete process and documentation used to get there.

Step Two: Collect and Analyze Data

The process to accomplish successful Completed Staff Work is grounded in how well you research the data that will be used to make your decision. This is the point at which successful people make their decisions based on actual and not perceived reality.

The importance of data collection and research to the success of your project cannot be understated. Data collection is the hard part of research and requires the utmost in discipline and tenacity. In a group process, there are always people who will persistently push the group to just make the decision. Frequently, you will hear this in the following form: "I don't know why we are wasting our time here; the proper decision is as plain as day."

The reason this part of the process is so hard is that we tend to try to find an easy way to make our decision and then move on to the next problem of the day. Formulating opinions on the right thing to do in any situation based on our perceptions often will prove counterproductive for realizing appropriate, long-term problem resolution.

> Formulating opinions on the right thing to do in any situation based on our perceptions often will prove counterproductive in realizing appropriate, long-term problem resolution.

Frequently, our perceptions regarding a situation are very different from the facts. This is primarily because we often do not use facts to form the basis of our decisions. The other problem is that as we grow in our roles as leaders, we develop prejudices regarding issues and tendencies to do things a certain way because that is the way we were successful on some other occasion. We should more appropriately be challenging ourselves to realize that most situations differ from each other. The facts regarding any two situations are almost never the same; for that reason, we need to commit to the hard work of data collection.

Until you get to the final stages of a decision-making process, experience and intuition are your enemies. Use them only when time is the critical factor in the decision-making process, or when all the facts lead to two or more possible recommendations.

Research Data and Information

The beginning of the data collection effort should start with basic inquiries related to the issue's history and background. One effective method for accomplishing this type of detective work is to trace the issue back to its origin. Consider first interviewing those who are currently or were previously affected by the issue.

Keep in mind that information obtained through interviews most likely will be based on the individuals' perceptions of the issue. Your focus is to work back to the origin of the issue, looking for those nuggets of information grounded in fact. Once you have developed a clear, accurate picture of the history of the problem, turn your attention to the current reality of the situation—a view of the situation that is not based on opinion and can be represented as fact. Write these facts down and look very closely to see if any of them need additional research to quantify, or raise additional questions.

Understanding the history of an issue and the current reality of your situation still leaves many unanswered questions. Looking over the information, look for gaps that may raise additional questions. Likewise, look for threads that tie details together and make it whole. Ultimately, your goal is to question everything until you have all the information needed to form a complete picture of the issue.

Involve Stakeholders

If you did a comprehensive job of identifying your stakeholders in Step One (and, more importantly, are using them to collect current reality data), they should already be intimately tied to your project. Now it is time to consider how involved you want each of the stakeholders to be in the next phases of the process.

Some stakeholders can be used to participate in discussions and be a part of the formal group. Others can be used as technical experts to answer questions pertaining to the data you discovered. Also remember to consider who will write the report and review the draft, if that is appropriate. The idea here is not to waste people's time but to gain the positive results of meaningful participation. At this point you should feel comfortable that all your stakeholders have been considered and have had time to participate and provide their input.

As mentioned earlier, influential people in your community called community issue leaders may be involved in social causes or be informal leaders for issues that affect the community's social life. In some cases, they may be heads of community civic organizations or leaders in your local chamber of commerce.

These folks may be important to the eventual implementation of your recommendation. Engaging them to assist in analyzing the research will give you a brief view of where they stand on the issue and whether you will have additional political problems implementing future recommendations.

Seek Customer Input

If you have not engaged your customers at this point, consider if their input is appropriate. Ask yourself what these customers want from this process and how you know they want it. Keep in mind that we are in the business of working for our customers and that allowing them input is critical to the eventual implementation of your recommendation.

The bottom line to this stakeholder and customer group issue is that you need to spend quality time considering their position and thoughts prior to developing and accessing your options, because at that point you will be committed and well past the point of no return.

The bulk of your data collection should be written and documented regardless of the eventual presentation. A detailed understanding of the facts, as outlined and presented in this step of the process, helps explain the methods, procedures, and history necessary to assist the decision-maker in understanding the full scope of your ultimate recommendation.

Step Three: Develop and Assess Options

The information you obtained in Step Two should have provided you with a clear picture of the issue. From there it should be easy to see what the root causes of the issue are and any disconnects keeping you from getting to your desired outcome.

Caution is suggested at this point. Often, people will collect information on the issue and then recommend their original perceived solution anyway. All your options, no matter how many are recommended, should be a direct result of the data you collected and not be influenced by emotion or your biases.

Note: The most effective way to analyze your current reality information is with a problem-solving and management tool called gap analysis, which was presented in the previous chapter. The gap analysis tool can be used to develop alternatives by identifying the gap between where you currently reside (current reality) and the result of where you want to be, represented as desired outcome. The answer to the issue or problem will be in recommendations that fill this gap and will allow you to get from current reality to the desired outcome.

Using Criteria to Develop a Final Recommendation

Remember the weighted criteria we discussed in chapter 21? Here is an opportunity to use the technique again. One of the best things about using the Completed Staff Work model is that the consistent factors you discuss become the criteria by which you can evaluate options. As an example, a criteria filter uses such issues as cost; budget impacts; political considerations; alignment with department vision, purpose, and values; as well as many other criteria to mathematically determine the best option.

As was previously discussed, this is done by first determining your criteria and then rating it as some percentage of 100%, depending on its relative importance when compared with other criteria. Using a standard scale from 1 to 5, with 1 being worst and 5 being best, you can then set up your formula to calculate the total based on the value you assigned the criterion and the percentage you established for its relative importance.

Problems can occur if individual members value criteria differently and, therefore, assign different percentages to the same criteria. When this happens, there is incongruence in the weight of the criteria between members; this causes confusion and inconsistency. It is not that they don't value the same things: they just don't see eye to eye, which leads to confusion, frustration, and sometimes even the perception of misconduct and manipulation on the part of the project members.

When all the options are run through the filter, you will be able to identify one option, based on the weighted criteria, as the best recommendation. Other options, though feasible, will not be as well recommended. The brilliance of this process is in the fact that you will be able to explain to the decision-maker the criteria used in assessing the options and why a particular recommendation was made over any other included in your report.

The criteria discussed in this section should be considered the minimum element set to use when assessing your alternatives. Whether you go to the trouble of developing a weighted criteria filter should be based on the complexity of your project and the overall potential impact any recommendation would have on the organization.

Develop Cost Options

The first criterion, and frequently the most disconcerting for fire chiefs, is cost. Often, alternatives come with varying levels of cost, depending on a variety of circumstances. For example, one recommendation may include buying a piece of equipment while others do not.

Each alternative and your final recommendation carry potential costs to the organization, whether they involve equipment, facilities, or personnel. Even policy changes frequently entail costs; knowing that up front is important to the decision-maker.

The cost of an alternative can cause a decision-maker to be led astray. Some folks consider the best alternative to be the one that does not cost anything, especially if the organization is strapped for money. Just keep in mind that your task is to make the best recommendation based on the criteria you used to rank the alternatives. Cost in and of itself should never be the lone criterion used to make the decision. If you are the project leader and the best alternative costs money, that is not your fault. Your efforts will be rewarded when you can demonstrate to the decision-maker the overall rationale for the decision.

Consider Budget Impacts

Related to the cost of a particular option is the effect it will have on your current and future budgets. Here, we are referring to the overall implication of the alternative, not the specifics of its cost.

To understand this, you must appreciate that decision-makers are often much more critical of recommendations that carry a recurring cost, which impact the budget year after year. It is especially important to consider this recurring impact regarding monetary fixes for problems that involve hiring personnel. As personnel gain in seniority and receive cost-of-living adjustments, not only do costs reoccur each year, but they could escalate well beyond the organization's ability to fund the recommendation over time.

Consider Policy Impacts

An essential consideration is how the alternatives mesh with current policy. The last thing you want to do is start one of those domino problems because your recommendation doesn't mesh with existing policy.

Regarding policy, the best analogy is to visualize your recommendation as a balloon filled with air. When you push on an inflated balloon, you displace a volume of air on one side with your finger, only for it to shift somewhere else in the balloon. Your job is to anticipate where the pressure will move and determine the effect it is going to have on the rest of the organization. Ultimately, make sure policy problems are addressed prior to (or as a part of) the overall recommendation.

Consider Noncost Impacts

Noncost impacts regarding Completed Staff Work can involve any number of considerations. You need to keep in mind the amount of change your organization can tolerate at one time as well as the prevailing culture.

As indicated in the discussion about criteria filters, political considerations are a big factor for the decision-maker. Determine if the political timing is correct and if the decision-maker has the political power or will to see the recommendation through to successful implementation.

Also consider the overall effects of politics in accessing your options. If you did a good job evaluating your enablers and restrainers (the forces that help you move toward your goal and those that hinder you from achieving your goal, respectively) earlier in the gap analysis, you will have a good idea of where potential political problems could arise. That said, politics in local government is a funny process that sometimes defies explanation. Do not be surprised if even the best recommendations—those that make the most sense—don't get approved because of politics. More on this subject later.

Consider Public Relations and Marketing Aspects

If your recommendation is going to have any effect on internal or external customers, good or bad, consider public relations. Making sure all stakeholders understand the need for implementing your recommendation is one of the keys to success. Open and honest communication is vital in public relations.

One of the critical things a decision-maker wants to know is how the recommendation will be viewed by whomever it affects. Make certain that you have considered these issues; at the very least, alert the decision-makers of any potential public relations issues.

Closely associated with public relations is marketing. In this case, successfully market your recommendation by using a defined communication process and your key stakeholders to spread the word about why the recommendation is the best for the organization. I have always felt it was in my best interest to float an idea out in the organization and then wait to take its temperature. The temperature in this case reflects the overall feeling of the organization and determines who the potential sabotaging members are (if any). Identifying these folks up front will allow you to do some intense lobbying prior to releasing the report.

Step Four: Recommendations and Draft Report

At this point, you will have run your alternatives through the filter and consider all the issues involving assessing the best option. You are now ready to start the final process of making a specific recommendation.

Seek Consensus of Stakeholders on Recommendations

There comes a time in all committee processes when the group needs to line up behind a particular recommendation. Evaluating all the facts using the information gathered in Steps One through Three, the group must come to some sort of consensus on what the final recommendation will be.

Obviously, not all group members will agree in all cases. You would hope that this would be the case after all your work and effort, but inevitably some members will view the value of the recommendation by the way it affects them personally, not by whether it is in the best interests of the organization.

If you are the project leader or even a member of the group, hold these members accountable by reminding them of your desired outcome statement. As the late leadership guru Dr. W. Edward Deming said, "Look at who is in the frame." Those members of your group who can see only themselves in the frame will never agree to a recommendation that does not serve their interests.

Select the Best Option

Your criteria should have guided you to the best option. Now it is only a formality to review the facts and draft your report. That said, take the time to go back and review your current reality information (the data) and make certain your final recommendation is based on facts and gets you to your desired outcome. The focus should be on how well the recommendation answers the organization's issue and serves the best interest of your customers.

Resolve Conflicts

Conflict resolution at this point should be minimal; however, certain members might not agree with the selection of the best option. In this case, see if there is a solution or compromise that addresses everyone's concerns and still provides the desired outcome.

In bad cases of conflict, project team members can take on the role of distracter or restrainer. Others who up to this point were fine may become bullies, getting angry because they see that things are not going their way. You may also encounter pressure exerted on team members by ranking officers of the group who use their position to influence others.

If you have members who, with the best interest of the organization in mind, just prefer one recommendation over the other, let them know it is okay to disagree but not to be disagreeable. This sort of situation does not normally occur, but if it does, give all members time to voice their opinion and then use the criteria filter to further guide the consensus process.

Legitimate conflicts can arise when members find that the data is incorrect or a mistake was made in calculating the results. In this situation, a full review and potential modification of the recommendation may be necessary. Make sure the incorrect findings are legitimate and not just a smokescreen put in place by distracters.

Consider Presentation Method

When choosing the most effective way to present your report, consider how you are viewed by the decision-maker. If you are a trusted and proven employee, you may only have to provide a bulleted list of information. If the decision-maker is not familiar with you and your issue, you may need a more thorough way to fill in the information gap.

Whatever you do, do not make the mistake of giving loads of data and graphs to a decision-maker who wants only an executive summary. Conversely, do not shortchange a decision-maker who likes to get into the facts and determine how you reached your conclusion.

Not only will you want to give some thought to how much information the decision-maker gets, but you will also want to decide the form in which that information will be provided. Some people like the full report along with an electronic presentation that highlights the important points. This way, if questions arise from the presentation, they can go back to the report for more information.

Consider also how visual the decision-maker is when it comes to comprehending information. For some folks, charts, graphs, and maps are the ticket. Others could care less about this flashy stuff and just want the facts.

No decision-maker only moderately concerned about your issue wants to read or suffer through a boring presentation. There is nothing wrong with asking the decision-maker(s) which type of presentation they prefer so you can be certain that you get the most out of the decision-maker's time.

All things considered, I have found it is always best to give the decision-maker a report even if you do not expect it to be read. Many times, a trusting decision-maker will never review your reams of information; however, that should not preclude you from providing it as a resource in case there is a question later. Finally, make sure your formatting is professional and all the words are spelled correctly. Nothing can ruin a presentation like misspelled or misused words.

Draft the Report

Your final piece of work will come in the form of a report. This report will speak to your professionalism and to the quality of your group's work. Make certain it is a quality work and represents everything you want it to be before it is handed off for review.

Along with checking the appearance of the report, make certain the report is complete and, above all, accurate. Regarding context and clarity, your goal should be to produce a report that any casual, or even uninformed, reader will understand. Ask yourself, if you were the decision-maker, would you sign off on the research and recommendations included? If your answer is yes, you are ready to release the draft for review.

Circulate Your Draft Report for Feedback

Up to this point, you have hopefully done a great job of communicating about your work throughout the organization. It is now time to pick out a select group of stakeholders for a more in-depth review of your document. Consider including some people in the review process who have an in-depth knowledge of your issue and some who do not have such a depth of knowledge. In this way, different viewpoints will be represented and there is a better chance of detecting something you may have overlooked because you were too close to the issue.

Make sure you give your reviewers plenty of time to look over the document and provide you with feedback. Do not rush this part of the process. More than anything else, you want feedback to be meaningful and to lend professionalism and accuracy to your final product.

When you receive the feedback, give it honest consideration, even if it is critical. Make sure you address deficiencies when mistakes are brought to your attention. Do not be one of those folks who asks for an opinion but doesn't really want it unless it is what you expected it to be in the first place.

Once you have made corrections and considered the feedback, have all project members sign the report. Group members' signatures are the seal of approval you give the decision-maker that this is truly Completed Staff Work and the results are the best possible at the time.

Step Five: Obtain Decision-Maker Approval

It is now showtime for your project, when all of your hard work starts to pay off. The report has been professionally put together; all that is left is the formality of the presentation. This is the scheduled time you have set aside for your group or the group's project leader to give the type of presentation you determined would be appropriate for the decision-maker.

Now is not the time to begin thinking about the presentation. Your previous homework relative to the decision-maker's learning style should have already been matched with the correct presentation method.

Some decision-makers like a brief in-person overview that details the finer points of the project and the rationale for the recommendation. This might include a cover page or an executive summary of the report, which covers the main points in a timely manner. If your decision-maker learns visually, your options are wide and varied. PowerPoint slides are popular, but whatever the medium, be comfortable using it and make certain it is appropriate for your audience.

If using electronic media for your presentation, make sure you show up early to ensure that everything is correctly functioning. Prepare for the presentation by delivering it before your staff or another group before the "big show."

Alert the Decision-Maker to Potential Unresolved Conflicts

Although they are most likely not a part of your final report, there may be times when you will have unresolved conflicts. This is especially true if you are dealing with a controversial subject or the affected persons are going to resist changes the recommendation will bring.

As a decision-maker, I always appreciate when my folks tell me if things could get rough. Don't set me up by painting a rosy picture so I respond favorably, only to find out later there is unrest or even resentment regarding the recommendation.

Alert the decision-maker to who is happy or not happy with the report and why. The decision-maker will also want to know to what degree the affected stakeholders will support the recommendation. Consider also advising the decision-maker of any minority opinion even when the recommendation is overwhelmingly supported.

More importantly, advise the decision-maker of any political strategies you used to seek input and formulate your report, and of any recommendations you have regarding the political strategies you feel will help ensure successful implementation.

Determine Council Involvement

In my city we have a city council. You may have a county board of supervisors or a strong mayor. Whichever the case, you must consider whether the report needs to go to the elected governing body for approval. Most of the time, this issue will already have been determined by this step because of the nature of the final product; however, your boss may want the elected body made aware of the report as a matter of courtesy.

Step Six: Present to Council (If Appropriate)

If the report does go to the next level, you will consider the presentation approach just as you did for the decision-maker. The higher you go, the higher the stakes are for your department, your boss, and your individual future.

Determine the Best Presentation Method

From this point on, the discussion concerning the presentation has less to do with what the product looks like than with how it can best be communicated. The process used will depend on the complexity, seriousness, or perhaps the political potential of the recommendation.

Dealing with multiple learning styles can be challenging. Often, you can be successful by presenting the information so that the visual folks get the gist of the situation without making it so visual that the analyzers in the audience are put off by the whole thing. As noted earlier, it is always better to provide the full report even if the decision-makers don't take the time to read it.

Another important note concerning political decision-makers is that they are often overwhelmed with written information that can be very technical in nature. As a rule of thumb, a political decision-maker needs to know only enough about your work to be able to explain to constituents how their tax money is being spent.

Position the Product

Selling any good idea is more than just good research. In fact, even if the research is the best in the world, a bad presentation could prevent the decision-makers from becoming engaged. They may come to see the whole matter merely as something they have to survive. Getting them excited about the potential of the research so they can accomplish some mutual goal is dependent on how the product is positioned prior to the presentation.

If a little informal "setting of the stage" is necessary in the process, consider meeting with council members individually or in groups of two, depending on their ideology or political persuasion. Never attempt to set the stage with two council members who see things differently or espouse opposite points of view most of the time. Also, remember that government officials are restricted by rules and regulations pertaining to the number of officials that can attend informal and formal staff presentations at a given time. In my city, for example, if more than two council members are together at any given time, they cannot conduct any city business.

Usually, it is best to frame informal discussions aimed at selling the product to elected officials around two aspects. First, always expand on how good the recommendation will make the elected officials look to their constituents. Second, if many unresolved conflicts or areas that could result in political backlash remain, alert them to this as well. One thing to remember here is not to be afraid that the recommendation may cause political unrest. All change causes some unrest. Just make certain your elected officials are not blindsided by stakeholders who want to sabotage your work and, subsequently, make the officials look bad in the process.

Publish the Report and Make the Presentation

If you have properly completed these steps, what happens because of your published work or the presentation(s) that results from it should come as no surprise. You have done your work, and the recommendation is aligned because of the criteria used to determine it was the best option. Stakeholders have been advised, conflicts were resolved and communicated, the stage has been set, and now all that is needed is a council decision in your favor.

Obtain Council Decision

Up to this point, you have put in a tremendous amount of effort to make sure your recommendation is in the best interest of all those involved. That is what this process is all about. What you must keep in mind is that even if you conduct the entire process perfectly, there is still a chance that you won't get a positive nod on the recommendation.

As an example, let's say that Council Members 1 and 2 just battled with 4 and 5 over an issue unrelated to yours. Tempers flared, and a close vote along party lines split the group and left half of them upset at the other half's vote.

Now, you come to give your presentation, and everything looks great, just like you envisioned during practice the night before. But it's just your luck that the winning side on the previous issue decides to get a little political mileage out of your presentation because they know it is great work and will eventually make them look good. They speak positively about you and your work before and after your presentation. Then comes commitment time, and you find that the support is clearly split along party lines—although this time the members supporting you find they are on the losing side.

The point to all this is twofold: If you have an option to move your presentation to a more favorable and less contentious day, do it. If that is not possible, then understand the decision had nothing to do with your work and everything to do with politics.

Step Seven: Implement Approved Decision

Regardless of the level at which an approved decision is made, this is not the time to rest on your laurels. Plenty of work still needs to be done to communicate the approved decision, distribute the report, and develop an implementation strategy.

Most of the time, a recommendation will cause some group of people to experience change. When this is the case, you will want to ensure your final report is communicated to those involved stakeholders. Just as you have done during the entire process, make sure all participants and stakeholders are briefed on the next steps in implementing the action.

Determine Implementation Strategy

Now you should go back and review the many previous steps involved in your process to see how many of them will be helpful in developing a strategy for implementing your recommendation. I assure you I am not trying to get you to conduct this process all over again. However, it may be worth considering certain steps in the process if implementation is going to be complex or involve multiple levels of your organization. Make certain at this point that your project doesn't fail because you didn't give any forethought to how it would be implemented.

What Hand Offs Need to Occur

It is unlikely the issue manager is going to be able to perform all of the steps of the implementation process. In these situations, it is appropriate to hand off certain tasks and responsibilities. Be very careful when handing off functions that may be crucial to implementation success. Always make sure that the delegated tasks are accompanied by specific instructions and clearly outlined expectations.

If multiple agencies or personnel are involved in the implementation process, it is wise for the issue manager to maintain the lead role. Remember, the issue manager is the person most familiar with the project and the one with the greatest stake in successfully implementing the recommendation.

Step Eight: Follow-Up

An all-too-familiar mistake is for the project leader to fade away and consider that the job is now complete. It is essential to keep in mind that approval for the recommendation is just a step toward the desired outcome. The real test of the research and recommendation that follows is whether it solves the problem by addressing its root cause.

Monitor Implementation

The project leader must be able to effectively monitor implementation activities. This may take the form of benchmarks that determine if progress is being made or just checking with stakeholders to make certain plans are progressing as expected.

Evaluate Implementation

Finally, the real benchmark of success will be if the recommendation solves the problem. This evaluation should involve a review of current reality to see if the desired outcome has been achieved.

In all occupations, the ability to effect positive change is derived from employees using methods that look at problems and then solving those problems for the betterment of internal and external customers. This requires reviewing and acting on the facts, leaving personal perceptions out of the problem-solving mix.

In the fire and EMS services, we usually discuss success as it relates to the first few minutes of an emergency. If we intervene in the situation in a positive manner during the critical first few minutes, we are generally successful. These types of situations require quick thinking, using mostly sketchy information and the cumulative experience acquired during similar situations.

Completed Staff Work is not a very effective tool for use during these types of situations; likewise, our fireground problem-solving methodologies are not the right tools when time is on our side. When we have time and the stakes are high regarding the organization's future direction, there is no excuse for not thinking through and addressing an issue accurately and completely.

Using the Completed Staff Work model as a guide helps you make the right moves for your organization while cutting down on the inefficiency caused by poor decision-making. This means not taking shortcuts but instead working through a process that looks at the entire issue, not just the parts you think are easy.

The future of our fire service depends on people who are willing to raise the professional bar, not lower it. Making decisions about our future is important, and firefighters who study the relevant issues that confront all our organizations owe it to each other to make the right moves.

Try the Completed Staff Work process the next time you are assigned to resolve an issue within your organization. I think you will be pleased with the results, and so will your leadership.

Culture and Organizational Change

True leadership is often when someone questions the social norm and goes against the grain if necessary.

Key Points

- Culture is made up of the collective experiences of employees.
- Employees are influenced by others to determine normalcy.
- Culture and organizational change are first and foremost individual.
- Culture and organizational change are a process of stages that can be anticipated.
- Your past experiences are not necessarily relevant to your current employees.

I vividly remember a skit on the old *Candid Camera* television show where three people walked on an elevator and, instead of turning around to face the door, they continued to face the back of the elevator. As the elevator moved to another floor, the doors opened and other people entered. Those entering the elevator would instinctively push the desired floor indicator and then turn and face the front doors. Eventually, feeling somewhat out of place, the person facing the front would turn in the same direction as the others and face the back. One by one they would enter and behave in the manner that was familiar to them and then turn around and behave like the others even though the behavior was foreign and uncomfortable.

So why is this important to you as a leader? In the example of the backward elevator riders, imagine how uncomfortable it was for new riders to enter and be confronted with a situation in which the behavior of the occupants was anything but normal. Even though they had never faced backward when riding an elevator, they were compelled each time to behave in the same manner as other riders.

Individuals, the kind who work for us, are very impressionable when it comes to pressure and what is considered normal behavior. If you take a motivated and enthusiastic employee and place them on the work unit with a supervisor with a poor attitude, they eventually start changing. As sure as the day is long, these enthusiastic employees give in to what is considered to be acceptable behavior on that work unit.

> Individuals, the kind who work for us, are very impressionable when it comes to pressure and what is considered normal behavior.

This is what makes changing the culture of organizations so difficult. When people are part of a group, they are influenced by one or more of the members of that group. Social scientists have proven time and time again that folks will behave in a similar fashion when in a group even when they do not know why they are acting that way. In some cases, members can be convinced to do things even when they know that they are not the correct actions to take and/or the actions are directly counter to their own personal value system.

Let us look at the example of firefighter safety, as this subject should be close to the hearts of all of us in the fire service. If your organization does not continually reinforce the importance of safety, or health and wellness for that matter, employees will not consider it to be important. When safety, health, and wellness are not at the forefront of your efforts, employees slowly stop putting an emphasis on them.

The individual collective experiences of your organization's members define what is acceptable behavior. When enough individuals start behaving in the same manner, and that behavior is considered normal, it is then defined as your organization's culture.

Eventually less attention to detail, whether it is compliance with policy or apathy about accident prevention, becomes the norm. You can train the hell out of new employees; however, just as in the case with the elevator, when you place them in an environment of people behaving differently than they were taught, they will eventually succumb to the group's behavior.

The dynamics of culture and organizational change occur at the individual level in your employees' personal experiences, are refined in their individual work units, and then are used collectively to form the basis of what becomes the culture of your organization.

I have seen some bad things happen to some very good leaders; however, it is infrequent at best that these folks did not have a hand in what occurred. These unfortunate leaders almost never directly cause whatever bad happens, but they fall victim to being responsible for the culture that allowed it in the first place. The culture of your organization reflects your leadership, and for that reason understanding how to manage and lead culture becomes paramount to your success.

Over the years, I have been involved in helping change or refine the culture of several organizations. In every instance I watched the same cycle and process unfold in a similar fashion. Understanding this will help you develop an appreciation for where you are in the cycle, prepare you for the next step, and for that matter make sure you get all the way through the cycle's many process points.

Let me bring your attention to figure 23-1. On the left side, notice the rows titled observations and actions. The observations row is further divided into process and phases. The actions row lists specifics regarding the corresponding actions of stakeholders (read, employees) and leadership (read, you and your staff). At this point, do not get upset that we separated employees and leadership on the chart even though all of us realize that everyone can be a leader in the organization. Rather, in this situation, and for the purposes of this discussion, it is

Culture & Organizational Change

OBSERVATIONS							
PROCESS	Push for Different Normal		New Expectations Established		New Normal		
PHASES	Recognition	Discovery	Resentment	Consequence			Acceptance
STAKEHOLDERS (ACTIONS)	Discontentment		Seek Place in New Way	Early Adapters / Fence Walkers / Late Adapters / Nonadapters	Boundaries Tested		Maturation
LEADERSHIP	Complacent	Ambivalent	Open to New Way	Establish Boundaries	Coach/ Mentor		Confident
CYCLE	UNHEALTHY		HEALTHY				

FIGURE 23-1. The Culture and Organizational Change Cycle

necessary to draw a distinction between these two components of an organization.

You will also notice that the Culture and Organizational Change Cycle has a left, unfavorable side and a right side that we want to always be dealing with. The point to this is that you want to always stay on the right side, and if you find yourself on the left, something has gone wrong—more on that in a minute.

Let us start our look at changing the culture of an organization by reviewing the process. The process follows a pattern composed of first a push for a different normal, then a period of new expectations established, and third settling into whatever has been defined as the new normal. That is not to say that each time an organization goes through a change it hits every aspect of the change process in the same way. For example, you could be in an organization that is apathetic regarding safety on the fireground. As the leader, you could begin the process of changing the culture to ensure there is a commitment to fireground safety and most likely hit all aspects of the process. Alternatively, you could have a devastating event like a firefighter fatality and go directly to a new normal, in effect skipping all of the organization processes of pushing for a different normal and establishing new expectations. I have long maintained that that there are only two things that can change an organization's culture: patterned behavior over time or a significant event.

So, let us start with the process element of "Push for a Different Normal." The biggest aha moment for a leader is that if you are the long-term leader of an organization and find yourself in need of a different normal, something has gone wrong. You may also find yourself in this part of the process when you are hired into an organization. Many times, when a leadership change has been initiated on behalf of an organization (read, somebody prior to you got fired or was forced out), you will find yourself in this part of the cycle.

The phases that exist in the push for a different normal involve recognition and discovery. In this regard either the employees or the organization's leadership recognize that something needs to be different and then start experimenting to discover what doing something different really means.

Let us say you go for your weekly meeting with the boss, and he tells you that he went by one of the stations last week for lunch and could not help but notice that several firefighters appeared overweight. He also noted in your yearly report an increase in the number of lost worker days related to stress, high blood pressure, and on-the-job injuries. This is an example of being in the recognition phase of the "Push for a Different Normal" process element of the change cycle.

Armed with this observation and the undeniable charts and graphs from your human resources department, you start the process to determine what is happening in your workforce. Of course, at this point you certainly wish you were out in front of the issue, but as is sometimes the case it is difficult to see the forest for the trees. This is an example of being in the discovery phase of the "Push for a Different Normal" process element of the change cycle.

So, what are your employees doing while all this is going on? In some cases, the employees themselves may be the ones pushing for a different normal. Imagine a long-time leader who has lost energy to lead and manage change. This is not an unusual situation for leaders coming to the end of their careers and may cause employees to push for change.

When there is a perception that things need to change, employees are experiencing discontentment. They will just plain not be happy with the status quo and will let their voices be heard through public displays of dissatisfaction in both mainstream and social media. In the case involving the wellness example, this could occur through recognition that as a department you are failing to meet, or even attempt to meet, national standards.

> It has been my experience that the one thing employees dislike more than change is no change at all.

It has been my experience that the one thing employees dislike more than change is no change at all. They also do not want to be considered as second rate when it comes to innovation, although the challenge of being innovative is difficult at best when considering the discipline on their behalf that is required to achieve that end.

If you are wondering how leadership could allow a situation like this to occur, it is because while the employees are displaying discontentment with the current situation, leadership was complacent and then ambivalent: complacent about recognizing a situation exists in the first place and then ambivalent in not caring how or why their organization is positioned concerning that issue.

Assuming you survived this push for a different normal and have done adequate inquiry into what needs to be done to get out of the situation, you will move into the "New Expectations Established" part of the process. This is where new procedures are outlined and different expectations are established for how the organization will behave in the new environment.

Using our wellness example, let us say for instance that as an organization you have decided that new fitness standards need to be in place. To accomplish this objective, a policy is created that outlines expectations regarding each member's daily physical fitness and yearly fitness assessment requirements. In addition to training your employees in exercise physiology and buying them workout equipment, you provide them with nutritional counseling. The desired outcome of the new expectations is a change in the culture of the organization to put great emphasis on health and wellness and, in doing so, reduce sick leave and lost workdays due to injury.

During the establishment of new expectations, you can take it to the bank that you will also encounter a resentment phase. Of course, not everyone will resent the new changes, as the proposal may affect everyone differently on a personal basis. In our wellness example, some of your more elite athletes and fitness buffs will be thrilled with the proposed changes. On the other hand, the overweight and less than healthy segment of the workforce will see the changes as challenging and burdensome.

You will also enter what I call the first part of the consequence phase. This phase overlaps into the "New Normal" process elements, as it will take employees time to understand the new expectations. The consequence phase is the phase when some employees learn that there is a cause-and-effect relationship in life's choices. In this case, they can choose to comply, and the consequence will be good evaluations and a healthy body. On the other hand, they can choose not to comply, and they will be disciplined and remain with an unhealthy body.

When any organizational change is implemented, employees will behave in one of several ways. Some of them will be early adapters and champion the cause. Identifying these folks on the front end of the change process is integral to your success, as they will eagerly adapt to the new way and serve as ambassadors for the proposed change when dealing with the more skeptical employees in the organization.

While early adapters are great, they are far less prevalent than your fence walkers. These are the folks that are waiting to see how much turmoil the change is going to create before committing. Most fence walkers will not actively oppose your change; they are just reluctant to engage on the forefront. They are also the employees who, for the most part, understand the greater need of the organization over their personal situation; however, they will remain noncommittal until the winds of success start blowing in your direction. As a side note, confident leadership can sway many fence walkers as they are only looking for you to tell them everything will be okay. The moral to this part of the story is that as a leader you need to be confident, enthusiastic, and focus on the positive attributes of the change. Remember those traits from chapter 6? That effort alone will go a long way in getting folks off the fence.

The late adapters are not going to commit to the change until there is no chance that the change will fail to meet its desired result. These folks are the ones you need to watch out for as they are fearful of change and will on occasion try and sabotage your efforts. This element of your organization is harmful, as they can, and will, use a system of misinformation to achieve their desired result.

Unfortunately, there will also be a segment of your workforce called non-adapters who will never buy into the proposed change. For these folks, the change becomes a personal battle with the leader that becomes less and less about what is being proposed and more and more about the person who is doing the proposing. The fact of the matter is that even these folks need to understand that as a leader it is your job to lead and they will be required to comply. When they do not commit, there will be consequences.

The consequences mentioned here are going to come because of your leadership team establishing boundaries for the new expectations. If you are the leader you will need to make certain that the new boundaries are clear and therefore the expectations are as well. Without this effort, the leadership team you are relying on to help the organization work through the change cycle will not have clear direction on how to manage the expectations of employees.

At some point in all this confusion and mess you will enter the new normal. I will caution you that getting here requires a tremendous amount of patience. In normal situations, outside of a significant event, the process needs to play itself out, and there is no way to shortcut the pain and work that is required to be successful.

As you enter the new normal you will also be experiencing the last part of the consequence phase. This is when the late adapters and nonadapters will test the established boundaries to see if you are really committed to the change. As the leader you need to remain steadfast in your commitment to your leadership team. It is through their efforts in coaching and mentoring that late adapters and nonadapters are dealt with.

Coming out the other side of all this gives you the realization that hard work and commitment mean something and can change your organization and the lives of the men and women who work there. This is a time when a new normal has been established and people have a hard time even contemplating that there was a previous way that you did business. It is in this acceptance phase that your stakeholders have matured and understand the need for the change and your leadership is confident in its ability to manage the workforce.

The bottom of figure 23–1 includes arrows that show the flow of the cycle. In every organization, change is constant and takes place at many different levels. You need to realize that at any given time you are at many places in the change cycle because many problems are being handled simultaneously. The point is that no matter what, you want to move to the right side because anything on the left side means someone took their eye off the ball.

The right side of our cycle clearly shows that it is natural and healthy for new expectations to be established in organizations. It is also healthy, and should be expected, that there will be resentment and consequence prior to acceptance of the new way. Our employees will be all over the place with their reaction to and acceptance of any change, which will be determined based on how it affects them personally. To remain effective at managing change, leadership needs to always be open to exploring new ways to carry on the organization's work. In doing so, changes will be introduced, and leadership will need to establish boundaries and then coach and mentor employees until they understand those boundaries.

No matter what happens, do not panic and give up. Organizational development and, in particular, managing changes in an organization's culture is messy. The point here is to figure out where you are and be prepared for the next step in the process.

The last bit of advice I will leave you with concerning organizational culture is that when it comes to change you should remember that your experience may not be relevant to your employees' current experience. What this means is that your collective experiences, and the effect they have on framing what you view as important, may not be the same as for many in your organization. This is especially true if you are brought in from the outside. The moral to the story is put your head down and keep running.

> Organizational development and, in particular, managing changes in an organization's culture is messy. The point here is to figure out where you are and be prepared for the next step in the process.

Culture and organizational change can be assessed as a part of a cycle that can to some degree be assessed and anticipated. Some aspects are on the unhealthy side of the process, while others are healthy. As a leader you should strive to be on the healthy side of the organizational change cycle.

Process elements of the cycle move in a linear line from a push for a change, to a state where new expectations have been established, and then to what can be considered a new normal.

Phases during the process include recognition, discovery, resentment, consequence, and acceptance. You should be mindful that resentment and consequence, while negative in presentation, are normal and should be considered a part of the change process.

During the phases of the organizational change process, employees will respond with discontentment and then seek a place in the new normal. From this point they will fall into one of four adaptation phases before testing boundaries and then finally maturing regarding the change.

Leadership in the cycle will be complacent or ambivalent in early stages and then, as process changes are sought, move toward being open to the new way, mentoring and coaching, and then being confident in the new normal.

24

The City Manager, Politics, and the Fire Chief

As the fire chief, you should be fully aware that every political win may also come with an equally powerful political loss at some future date.

Key Points

- City management and fire chiefs have differing responsibilities.
- City management must balance and respond to issues from many sources.
- City management seeks collective power of elected officials.
- Working with elected officials requires year-round engagement and effective communication.
- Complexities of government services must all fit together to benefit the community.
- The responsibility of elected officials is to represent their constituents.
- Elected officials are motivated by service, ego, and reelection.
- Political wins are often followed by losses.
- The fire chief's role is to provide expertise and work with government functions.

The Dark Side

Yes, I have spent time on the dark side. You know what I mean: up on the hill, the glass tower, downtown, that place where administrators and managers do everything possible to make fire chiefs' lives miserable. Occasionally, we fire chiefs get invited to these ivory towers of executive decision-making for meetings and such, but we feel uncomfortable and out of place as the broader path of the discussion seems to start with "How are you doing?" and end somehow with "What can you do without?"

I spent time in many meetings as a county manager and was always surprised by the kinds of things we discussed. For the most part, time was spent trying to balance the needs of the entire community against those folks with special interest groups or politically connected citizens who could advocate for one thing or another. Perhaps the most important thing we did do was try to look out for the people who could not, or would not, advocate for themselves yet needed the kinds of services that would benefit the community at large and would add to the collective long-term well-being of the community.

As a lifelong firefighter, I must admit that for the most part I enjoyed my time as a county executive. I learned a great deal about the fabric of the community and even more about what it takes to create and maintain a healthy community. I also spent a great deal of my time working for the interests of my department heads. I worked with the very kind of department heads that I once was as a fire chief—technically proficient, extremely committed, and emotionally attached to achieving the organization's mission.

At the end of the day, I guess bilateral sewer lines, rezoning requests, board agendas, and endless nights of board meetings just got the best of me. Sewage, water, roads, and planning can be exciting, but as we all know they are not nearly as exciting as being a firefighter. I eventually moved on to another opportunity to lead a fire department. That said, I do think my experiences on the other side were well worth the time spent, and I have carried them with me.

With that as a background, I thought I would share with you what it is like to be in the role of the community executive. My hope is that you will gain perspective on the unique challenges of the position and then have open dialogue about how you and your manager can best work together. If not the former, you could instead empathize with each other, cry a little bit, and then go get a beer.

So, let me share with you the perspective of Mr. City Manager.

Good Afternoon, Mr. Fire Chief

My world is filled with meetings that start very early in the morning and continue until late in the evening. When I am not in meetings, I am on the phone, which seems to provide me with an endless supply of citizens who all think their issue is the only one I should be working on: pigs in the wrong zoning districts, property line issues involving fallen trees, sewer lift station malfunctions that result in someone's hallway not smelling at all like last night's dinner, squirrels in peoples attics, developers convinced you have it in for them, citizens who scrutinize every tax dollar spent and remain convinced you can run a very complex municipal operation with fewer people.

During the day, my every move, including emails, cell phone, and landline records are open and fair game to be scrutinized by the public. You would think that I would be the one most able to control my schedule, but it turns out that is not the case. For me to save time on my schedule, I must block out hours with words like "planning time" and "reserve for personal time." Even then it seems that I move from one thing to the next with no two issues being even remotely the same.

I work for seven people who all have a different opinion about what kind of community we live in and what kinds of things are important to the community. Each one of them has a different learning style, and therefore no one method of communication is the most effective for all of them. They also have diverse value systems, which means that our efforts to properly govern are based on a balance of their individual values and choices. What may be perfectly acceptable for some of my bosses may completely violate the value system of one of their peers.

I am in almost constant conversation with each of my bosses because they feed me an endless list of critical issues as are brought to them by the citizens they serve. Keep in mind that in a community of 100,000 citizens, four phone calls to the mayor equates to a crisis of major proportion. In addition, fixing any one of these issues may or may not be the most critical issue for one of their peers. In fact, depending on the politics of the issue, they may indeed be upset at me for finding a resolution.

The power of my bosses comes from the collective power of four of them: any four of them, regardless of their political affiliation. For this reason, I try to keep all of them in the collective know. I strive every day to make sure that when one of them knows something all of them are provided equal access to that same information. This is very important and very time consuming.

Each day I come to work I can choose to lead my elected bosses or let them lead me. To make matters even more complicated, any given issue, day, week, or year may need a combination of these two approaches. To some employees and citizens, it may look like I am wavering in my leadership style. That really could not be farther from the truth; the most successful management process is like an ongoing environmental scan to assess risk versus benefit on any given issue.

What you need to realize is that successful leadership in public government is only possible when elected officials, who may be partially motivated by getting reelected, helping their constituents, or walking the political party line, are encouraged and guided to make the right choices for the citizens. In this type of situation, I need to walk a fine line in leading the group. If I appear too aggressive in leading the elected body, they will feel bullied or, even worse, may appear weak to the citizens. If I do not lead the elected body, they may lose direction and not be able to align their efforts with current and future needs. The best I can hope for is that the elected body has a vision for the community and a strategic plan that can accomplish that vision.

> *Fire chiefs, especially those that have risen through the ranks in one community, need to be able to make the transition from rank and file to management and to understand both the sacrifices and opportunities that brings. The chief needs to understand the broader organizational responsibilities that come with the position and to be able to cope with the difficult situations that occur with personnel.*
>
> —Kim Payne, City Manager, retired
> Lynchburg, VA

Three things motivate elected officials: public service, ego (this puts them in charge), and reelection. That said, most of them are good people who want to do great things for their community. Each of them is a leader in the community and, at the lowest levels of elected public service, not paid very much for their time. For this reason, we can never underestimate how important ego is as a driver of their decision-making.

Speaking of reelection, the folks I work for are constantly running for office. Even when Election Day is months or even years away, they will be running for office. They may tell you that they are thinking of the citizens first, but in most cases, their real considerations are somewhere in between that and how the decision will affect their potential to be reelected. Do not get me started on how hard it is to govern during a time I call the "crazy season." The crazy season is that time leading up to Election Day when no one knows what individual community crisis will happen from one day to the next.

I cannot afford to waste time in meetings, and sometimes this makes it appear that I don't care or am not sensitive to your needs. That could not be farther from the truth, but your needs are not the only ones I need to be sensitive to. You can take some comfort in the fact that in general I trust you because you are a technical expert in your field. That is why I may be indifferent at times to your requests that seem to only serve one aspect of our community. It may be the right thing to do, but my job is to figure out the cost of that right thing when compared to other very important right things that need to be done. By the way, you fire guys in general do not cause me nearly as much pain as the "boys in blue."

Speaking of employees, I have a lot of them. Guess what, they all provide some service they consider to be just as valuable as the service you provide. They also run the gamut when it comes to education and pay. They are all talented, and my focus is always on how to value what each of them brings to the organization and community.

The high level of respect for your profession affords you a higher platform from which to communicate your issues. For that reason, it infuriates me when you or some of your folks abuse this preferred treatment at the detriment of the remaining workforce. How hard do you think it is for the accountant in finance to compete with what you do? It will help me if you advocate for all the aspects of our workforce that need to be in place for you to be successful.

With all that said, my job is an extremely rewarding one. When I review what is done in our community to make it a better place for people to live, work, and visit, I have the satisfaction of knowing that my ability to herd cats played some small part in the effort. At the end of the day, I can look back and note a park, road, fire station, school, or library as one of the many positive attributes I made come to life in our community.

Elected Officials

Since I spent a great deal of time whining—I mean, explaining—to you about the trials and tribulations of working with elected officials, I thought it would be appropriate for me to tell you first how you can work with them to help me. If both of us work at refining our skills working with elected officials, the two of us most likely will not get sideways with each other during the process.

The first thing you should understand is that politicians want you to be involved all year round. Do not just show up in April or May of every year to talk about your budget. This means you need to attend council meetings from time to time, even when you do not have anything specific on the agenda for your department. This shows them that you are interested in the broader context of governing and allows you to learn their specific "hot key" buttons and special interests.

When I started in this business, the only way to get elected to council was to first serve on the planning commission or some other board. These days, citizens are running and being selected on single-issue platforms, so many of them lack a comprehensive understanding of local government operations. Many of them, especially newly elected officials, do not have a clue how complex local government is or the wide array of services all of our departments provide. For that reason alone, you need to take extra care to determine the extent to which a newly elected official has been exposed to your operations.

As the chief, you should take great pains to educate politicians about what it is you do, and while doing so, do not be afraid to blow your own horn. Elected officials take great pride in discovering and then telling their constituents what they are getting for their money. When there is an opportunity to keep them informed of your issues and how they relate to the citizens, do so by inviting them to your various meetings. It also does not hurt to occasionally ask what it is you can do for them. During all this formal and informal communication, keep in mind that it is my expectation you are political but not partisan. Treat each of them the same. More on that in a moment.

The fire chief must be confident and work well within the community at large. They are uniquely positioned to be leaders in the community and the organization if they choose to take on such missions.

—Rhu Harris, County Administrator, retired
Hanover County, VA

When it comes to elected officials, I expect you to answer their questions and address their concerns in a timely manner. If one member asks you a question, answer that one person. If that one person asks you a question in an open meeting involving other council members, make sure all of them get the information. In the process, if you think something is going to smell bad, make sure I know what it smells like before they do.

Make sure you keep our elected officials informed of what you are doing. Remember that I am busy and may easily dismiss something that your folks do because I am not as close to it as you. In most cases, I want you to notify politicians of your activities ASAP, sometimes immediately if they will be viewed in the media in something other than a positive light. Call me and let us talk through the issues so that we can work out the best approach for keeping them informed. If you have a question about the right thing to do under these circumstances, let your stomach advise you. You know what I mean: when you have that nagging sick feeling in your stomach that something is way wrong, that is when I need some warning time.

You should also not be afraid to invite me or other elected officials to your post-incident briefings. For the most part, they are even more clueless than me on what constitutes a successful fire. How could they know, for that matter? You know as well as I do fires that are stopped very quickly are attributed to an excellent knockdown, whereas fires that reduce buildings to foundations are considered terrible fires in which your gallant efforts somehow saved us from a community conflagration.

When you are dealing with elected officials, always consider them to be the voice of the people. If you can come to recognize that in many cases elected officials are making decisions based on constituent approval and not necessarily even their own value systems, you will come closer to understanding their perspective. Growing relations between the elected officials and you as chief requires you to work within the system of politics and always have consideration for their unique perspective.

To grow relations between you and the elected officials, always try to maintain a consistent and relevant dialogue. You should be political, engaged, focused, and determined, but do not expect to win every time. Remember that they may say something in a meeting that is intended for some political purpose other than to get you enraged. In the end, you need to view our elected officials as a small but also essential part of your fire protection system. In any case, do not write off those who oppose you, as the political winds and tides of time change very quickly. You could very well find out that your adversary today is your best advocate tomorrow.

Realize that it is the elected official who determines the acceptable level of risk for the community, and your job is to advise them as to what is appropriate. When there is disagreement in what you think you need and what they are willing to fund, it is your job to stand up for the citizen's safety. In other words, safety is your job and funding for safety is theirs. When elected officials do not do what you perceive as the right thing, do not fall on the sword to make a statement. Your folks need you the most during the tough times, and your career-ending comment at the council meeting is not going to serve them or the citizens very well.

When you are making presentations, try to present your information in several different ways. Doing this will provide great options when it comes to varying learning styles. For the fact freaks, give them the details and exhaustive research. For the visual folks, give then the overview with charts and stories. The bottom line is that you need to know your stuff and appeal to many different types of personalities and learning styles.

I also want you to know that, in general, elected officials and city managers are looking for a reason to say yes instead of no. To get the process working in your favor, show how what you propose is good for the citizens and that the benefits will in fact clearly outweigh the costs.

Whatever you do, please do not be emotional. This is a business, and they are not nearly as close to the real world as you. When you get emotional, your arguments are hard for them to read because they did not come up in a system built of emotion. For instance, I am a numbers guy; your red face means nothing to me. Save your emotional outburst for the locker room of your fantasy football team.

Politics

A strong fire chief has the leadership skills to continuously evaluate and improve how things are done within the department, between departments, and within the community.

—Tony Gardner
University of Virginia
Darden School of Leadership

Many elected officials view their main job to be protecting the public from the ill will of the bureaucrats. They are elected on a platform of harnessing the bureaucratic beasts because they believe that bureaucracies are created to self-perpetuate and grow larger. In some cases, they are right on the money; however, the fact remains that these two philosophies are in direct conflict with one another.

As indicated previously, try to find alignment by framing your issues as public policy issues. In doing so, the elected leader will be more apt to agree and help because the solution will accomplish something larger in the community than your individual departmental objective. In all cases, give them bullets and, more preferably, options that they can pick from to achieve the stated goals.

As the fire chief you are usually insulated or separated by a few levels from the real politics—that is, unless you fail to understand what I just said motivates them. This means that the manager or mayor is your buffer between the elected official and the bureaucracy. In all cases, the fire chief needs to understand the political sensitivity of a given situation and then be careful not to play politics. If you play politics and lose, you may find yourself on the outside looking in when the manager's support is needed.

Success at the political level is usually dictated by current community hot button topics or the crisis of the day. To be successful, you should look for alignment opportunities that exist between what the elected official is trying to accomplish and what support your department can provide to help reach that end.

When it comes to getting back to them with information, you should be aware that in some cases the political element does not want you to define the problem, they just want the problem to go away. That means all your root cause analysis talk will just be mumbo jumbo since they are in most cases only responding to an individual request for action. Finding the root cause may not be nearly as important as making the constituent happy.

It is most imperative that you do your homework before stepping out on your own when an issue could be controversial. Success in politically sensitive issues will be dependent on the support you have from both internal and external resources. On more than one occasion, I have witnessed the city manager with their head down at a council meeting when a department head really needed support. In most cases, their head is down because they are smart enough to realize that you do not have the support necessary to carry the weight of the situation. Remember, the manager is your link between politics and the bureaucracy of the government organization, and they need a job as well.

Most of the time, fire chiefs get in trouble because they come from a black-and-white world, where there appears to be some sort of a morally or ethically right and wrong side of every issue. Remember, successful politics is never about a clear winner or loser but rather almost always about compromise. In contentious situations where matters are resolved by compromise, everyone better walk away from the table with an equally unpleasant taste in their mouth.

That means you need to compromise such that no one gets everything they wanted but can still walk away from the table with their head held high. If it ever turns out that you find yourself on the winning side of a political issue, you can bet your bottom dollar that eventually you will find yourself on the losing end of another. Someone once told me that for every political win there is a political loss. If that is true, and I do believe it is, then at best one will only break even over the long haul. I do not know about you, but I did not get in this business to break even; I want to be successful. Do not create a win-lose scenario for the politician.

Let's talk about the fire chief who finally decides that he has had enough regarding the department's staffing and decides, *I want that fourth man on the engine, and I am going to politic for it as hard as I can.* So, politic he does. Community support builds as presentations are given to community leaders on the positive attributes of four-person staffing. Increased effectiveness and firefighter safety seem to be the compelling argument that creates citizen support for the initiative.

> *The effective fire chief needs to always maintain a positive outlook. Things are not always going to go your way regarding wants and needs. During these times it is critical the fire chief understand that the men and women in the department take their emotional clues based on the chief's behavior.*
>
> —Sterling Cheatham, City Manager, retired
> Wilmington, NC

In times other than a recession or declining revenues, the community will generally support the good guys. So, what is the politician going to do in this situation? In most situations like these the fire chief will prevail, as no politician wants to be seen as the one who contributed to a firefighter death or serious injury. In the end, all is good with the world as firefighters are singing the chief's praises. Hooray for the chief.

However, the future will surely deal this chief an equally as undesirable political loss. Remember, the politician needs to get elected, and for that they need votes. You don't provide votes, but you can create a situation in which elected officials look bad and lose them. In this game, you will always lose. Unfortunately, many new fire chiefs do not catch on to this compromise thing and fall victim to the one-for-one political wins and losses game.

The only way around this is to win your battles incrementally. Incremental politics is the art of understanding where it is you want to go over the long term then having the patience to accomplish your actions in small steps. You effectively build a foundation for successful change by introducing process elements at times when the organization can tolerate the change.

All is not lost though, as there are a few things that will help you achieve success. The first is to keep me briefed on your issues, especially those that could be viewed as politically volatile. Ask me for my support but do not frame me on the morally wrong end of the issue if I think that the timing is not right. When the timing is right, I will instruct you to brief council about the issue, and in doing so I would like you to educate them about the issue by speaking only to the facts. Resist the urge to offer your opinion or perception unless it is specially asked for.

I would advise you to run away from issues in which you are involved where there will be a political loser. There is a difference in you operating in a political environment and you being perceived as politically partisan. Your success may well be in how you manage these situations when they occur. When you find yourself in a position to support one council member but it may come at the detriment of another, you need to let me know. I am here to shield you from this type of situation and can do so if it occurs prior to you being too far out front.

In all cases, please do not get me caught in a situation in which you have advocated politically for something that I will then be forced to implement at the detriment of the other services we provide. You will not end up winning that one, if you know what I mean.

Working with Me

The thing I most need you to understand is that you are a part of the greater whole and only one piece of the community puzzle. I need you to understand how fire, rescue, and emergency services fit into the larger picture of what is important to the success of the community. As a part of my leadership team, it is imperative you understand that not everyone, even upper management, will get the priority of the fire department, so please consider balancing the needs of the fire department with the greater needs of the community in general. The fire department does not stand alone or above all others with respect to services needed in our community.

You should be proactive in solving organizational problems without specific direction from me but at the same time have an appreciation for keeping me informed about the right issues. That means the good, the bad, and the ugly. You should not be so independent that I need to search for general information about your department's activities and direction. Keep in mind that the longer you wait to keep me informed about issues and initiatives, the harder it will be to get me up to speed when issues arise. The rule is "no surprises."

Like almost everything else in life, people run into problems when they compromise reasonable standards and values. As the fire chief I expect you to have a strong personal value set as a baseline. Trust and integrity are the foundation for a strong relationship between a manager and the fire chief. You need to feel comfortable questioning me, and conversely, I need to be confident enough to respect your questions.

My way or the highway attitude does not work today. The fire chief needs to be a strong leader, but confident enough to realize that they do not have all the right answers. They need to be willing to call on the strengths of others in and outside the fire department even if they are in the lowest levels of the organization.

—Kurt Carroll, Village Manager
Village of Shorewood, IL

Pay attention to community issues other than your own. I am not only paying you for your technical expertise in firefighting but also your expertise as a leader. You need to state your opinion on what makes this a great community and contribute to the conversation. Do not be a wallflower until someone mentions increasing the lifespan of your apparatus replacement cycle. If you stay in the ballgame year-round, I will appreciate your unique perspective. Just remember your perspective is one of many that I consider.

You also need to be involved in our long-range planning. Do not let me think things are all well and fine and then wake up one day and tell me you need another fire station or medic unit. I need you to use all the analytical tools at your disposal to provide relevant data on what we need to serve this community. When you are considering long-range planning, think about how other community services can be helped by your service platform.

Respond to citizen emergency needs in a fiscally responsible manner. Just like your own finances, there is not an endless supply of money. For this reason, anything we do to grow our system or provide a higher level of service needs to be compared with the overall system cost. That is especially important if we need to rob Peter to pay Paul to accomplish your objectives.

I need you to be innovative and never satisfied with the status quo. Be open to change, and not just the type of incremental change we talked about earlier. Sometimes we need to build incrementally on what is currently being done, and at other times wholesale change is required. As the technical expert, I need you to be open to both scenarios and advise me properly if and when either of them is required.

I expect you as the fire chief to be technically competent in your discipline but also knowledgeable in budgeting, human resources, and community planning. You are only as important to me and this organization as far as you are willing to understand the other aspects of our service delivery system.

As fire chief, you should view issues as part of our management team. Each of our department's problems is unique, and therefore everyone's perspective is important and critical to the conversation. It is most important to our success as an organization that you establish and keep clear, open lines of communication between you, me, elected officials, employees, and the other department heads.

Lastly, I need for you to see your personal role within and beyond your own department as a coach, mentor, and developer of people who thrive and contribute to the department and the larger organization. Be a coleader in the community by understanding organization-wide challenges and embracing the responsibility of coleading the organization with me, the leadership team, and council.

As the fire chief, you come with built-in credibility. The public loves you, and city management respects you. When compared with many of your peers you are in an envious position.

As a department head, you have a responsibility to understand the complexities of your manager and the elected officials' duties. They are broad and are at times a compromise of values as presented by the citizens we serve. While you need to protect and defend your proverbial silo, you also need to respect the fact that yours is not the only silo in the field.

The fire chief needs to understand that city management is balancing the needs of the citizens with the aspirations of the elected officials that were put there to make them happen. Many of these aspirations are different and sometimes conflict with yours and those of the other elected members.

You should remember that the balance of power in every local government rest in votes when it comes to any specific issue. In a seven-member council system issues can turn on a 4 to 3 vote. Additionally, please respect the fact I can be fired when the majority on the governing body agree they would prefer a change in leadership.

Fire chiefs have a role in working with elected officials. However, that cannot come at the expense of the manager. We must work to keep leadership advised of potential detrimental issues because we are a team, and we both want to serve the citizens to the best degree possible.

25

Surviving Leadership

*It doesn't always work out like a fairy tale. When
it doesn't, always take the high road.*

Key Points

- Adapt your management of relationships based on style.
- Treat working with people who do not complement your style as a learning opportunity.
- Practice learning from both positive and negative situations.
- Happiness is a personal responsibility.
- People have the right to be who they are.
- Understanding yourself is a first step in understanding others.
- Manage expectations when styles conflict.
- Change your perspective and focus.

As a young fire captain working in the training division, I would on occasion take lunch while watching the wildlife in our training center pond. I found this a very relaxing respite from the ringing phones, course-learning objectives, and firefighter recruits.

During the three years I worked in training, I watched the same set of ducks return to our pond. Each year after their arrival we would have a pond full of little ducklings with the mother duck leading them from one end of the pond to the other. One problem for our visiting ducks was that they were not the only creatures living in the pond. Turtles, the big snapping variety, also found our pond a nice environment in which to raise their young. These turtles seemed to especially like little ducklings!

Each year we would watch as one by one the little ducklings fell victim to the snapping turtles. This process would continue over the course of a week, until there were no more little ducklings. Notwithstanding their lack of success for raising little ducklings, the same set of ducks would return the following year, and the process would repeat itself.

In the world of leadership and followership, this story contains a powerful lesson. First and foremost, some of us are the ducklings in our organizations and may in fact be working for turtles. Turtles, in this case, are disguised as leaders who stay hidden under the water and pluck unsuspecting ducklings off one by one until everyone remaining in the organization looks and acts like a turtle.

> In the world of leadership and followership, this story contains a powerful lesson. First and foremost, some of us are the ducklings in our organizations and may in fact be working for turtles.

Surviving as something other than the most popular or powerful species in the organizational pond can be tough, in the sense that it is hard to maintain a good attitude and stay focused when you are fearful something is lurking under the surface just waiting to pull you and whoever looks like you right under the water. Rest easy though: there are some things ducklings can do to survive the turtles and live harmoniously in the pond.

The Situation

All of us will at some point in our career find ourselves working for someone who just does not complement our leadership style. Perhaps you have a high interpersonal relationship need, and you are stuck working for a boss who doesn't even know your name and who only cares about the bottom-line results of your work. Maybe your boss has a high interpersonal relationship need and all you want to do is be left alone to do your job.

Whatever the reason for the conflict, there can be no more frustrating a relationship than working for someone who does not share your values or concerns. I believe this to be of relevance for fire chiefs and firefighters because we grow up in our jobs fueled by emotion and compassion. Unfortunately, we sometimes end up working for people who were reared in a system steeped in business model practices and political volleying.

Regardless of your situation, if you do not spend time cultivating an acceptable working arrangement between you and your boss, life is going to be miserable, particularly if you're a duckling and your boss likes working with turtles.

Change Your Perspective of the Situation

Perhaps the most profound thing ever taught to me was that we have an opportunity to learn from every situation if we choose. For instance, we can choose to be unhappy because we work for a poor leader, or we can use the opportunity to learn what not to do when we are the leader and presented with similar situations.

I remember sitting in meetings with my boss and other peers early in my career and thinking, *This guy doesn't have a clue.* The concept that there was a gap between his leadership style and the needs of his employees simply eluded him.

During these meetings I would find myself scribbling notes in my planner and thinking to myself that I would never want to do this when I had an opportunity to lead. As a matter of fact, writing down the bad traits and miscues of this leader helped reinforce the things I wanted to work on to become a more effective leader.

At some point I remember becoming so frustrated with this boss I sought the counsel and guidance of a close friend. I remember trying very hard to understand why I was feeling so unhappy about my situation and how I could possibly turn things around. As I related the stories of this boss and his behavior to my friend, I referred to my book and the notes I had taken. I really did not want to miss a thing while trying to justify my poor attitude about his leadership.

Out the blue, my wise friend said to me, "Every situation in life, good and bad, presents us with opportunities to learn and become better people." It was then I realized that by concentrating on this leader's negative behavior I was making myself feel bad. By turning that around and viewing it as a learning opportunity, I subsequently benefited from the relationship.

Happiness

If changing our perspective is the first lesson to learn in working with people we do not understand or care for, closely behind is our understanding of happiness.

As I mentioned in the chapter on traits, understanding that happiness is a personal responsibility that we have total control of is essential to surviving in any relationship.

Merriam-Webster's defines happiness as "a state of well-being and contentment." From the very definition, it is easily noted that well-being and contentment are developed from personal feelings that are intimately internal to us and us alone.

It greatly irks me when I hear people say that someone else does not make them happy. What these people do not understand is that happiness in life is internal, something that they alone are responsible for and that only they have control over.

It is never anyone's responsibility, least of all your boss, to make you happy with your job. That said, a good boss will try to create an environment where you are happy and can prosper. It is in their best interest for you to be happy because you will be more productive in your work.

> Perhaps the most important thing for any employee to understand regarding their boss's leadership style is that they have the absolute right to be exactly like they are. You have no right to expect them to be like you, nor should you want them to be.

Understanding the personal responsibility for happiness is one thing; surviving in your job while being unhappy is quite another. The important thing to remember is that it does not matter whether you are a duck or a turtle; you should focus on the root cause of the unhappiness.

Your task, as an employee or a boss, is to recognize the real root cause and avoid trying to convince people they should not be unhappy. If your people are fearful, lonely, or insecure, telling them that happiness is a personal responsibility will only result in more disillusioned and distant employees.

Their Right to Be How They Want

Perhaps the most important thing for any employee to understand regarding their boss's leadership style is that they have the absolute right to be exactly like they are. You have no right to expect them to be like you, nor should you want them to be.

In our personal lives, we have choices concerning who we like and do not like. Separating ourselves from people who do not add value to our personal lives is much easier than doing so in our professional lives. Plus, as you separate yourself from difficult situations each time they occur, you miss valuable opportunities to learn.

While one could argue that none of us are indentured servants to our organizations or its leaders, separating ourselves is usually a bit more complex. Retirement systems, kids in school, or any number of other events in our lives may result in us having to stick it out until the boss changes.

Once you have an appreciation for someone and how they view the world, it is usually easier to find ways to work with them, like your weird uncle who comes to holiday dinners and gets drunk and obnoxious. You really wish he weren't that way, but your solution to alleviate the problem involves removing yourself or him from the situation. In the case of your boss, who do you think will end up being removed?

Do Not Have an Expectation They Will Change

Everyone we work with is the way they are because they have decided to be that way. You will do well to understand this and just get on with the work of the organization because at this point in your career, or your boss's career, it may be very unlikely that either of you will change your leadership styles dramatically. You should not have the expectation they will change or that you can mentor them to see things your way.

This is not to say you and your boss cannot refine your styles to become more effective in leading others; however, wholesale changes in style are just not going to happen. Remember, everyone's leadership style is developed over time and is like the culture of an organization.

Understanding Yourself First

Before you can possibly understand why you are unhappy with the leadership of others, you need a profound understanding of your own view of leadership. What I mean is an honest assessment of your interpersonal relationship needs, leadership strengths, weaknesses, and tendencies.

There are several ways you can assess your view of the world, learning style, and personality. For instance, most people have taken the Myers-Briggs learning style test. Still other tests can determine your relationship tendencies regarding personal wants and relationship needs.

Perhaps the best assessment of yourself will come from your family, friends, and coworkers because they are the ones closest to you. Ask them to help you by conveying their perception of your abilities, both technically and emotionally.

If they are real friends, they will be honest in their assessment of you, and it will help paint a picture of what kind of person you are and what they believe floats your boat. Compare this information to how you perceive yourself, and you will be well on the way to understanding who you are and how you affect others.

Understanding Your Boss

After you understand your wants and needs, it is possible to begin the assessment of another person. You should begin this assessment by researching their background. Individual values and beliefs represent the sum of life experiences and are reflected in our behavior toward others. The aggregate of all this is our leadership style.

For instance, if your boss has military experience, they may have a high regard for commitment and discipline. If they have lived the life of hard knocks, they may have no tolerance for anything less than a very intense work ethic. On the other hand, if they grew up in a family-oriented environment or they have suffered losses of close family members, their value may be high on personal relationships and family.

> It is important to remember all bosses care about something. Your success in working with them is dependent on figuring out what is important and why it drives their leadership style.

Spend a little time doing research on your boss's background and how their life and job experience developed. It is important to remember all bosses care about something. Your success in working with them is dependent on figuring out what is important and why it drives their leadership style.

Individual Expectations

One of the most important things you can do to bridge any gap you have with your boss is to spend time with them discussing personal expectations the same way you do performance expectations. This may be very difficult for those of you who are already in leadership roles.

Chiefs and city managers, having already developed egos, may have difficulty admitting relationship-based feelings, wants, and needs. The problem is most of us grew up in systems focusing on hard stuff like performance expectations. The softer side of business, which is grounded in relationships and feelings, is very difficult for some people to confront.

Setting your position and ego aside and having an honest conversation about personal needs will help you and your boss come to an understanding about how the two of you can help each other. If you can find out what the boss needs to make them look good in their boss's eyes and develop a shared strategy for achieving that goal, you will begin the steps of building a better relationship. If nothing else, you will find out those things that really matter in your relationship with each other.

Take Advantage of Their Style

If you find the work arrangement and relationship between you and your boss acceptable, you should concentrate on complementing their style for the purpose of achieving your objectives. When I say, "take advantage of," it is not intended to be negative or underhanded. You should never do things for the wrong reasons or violate your organizational or personal values.

If your boss values knowing the details when you make a request, you need to do your homework and provide thorough information. If you do not, your request will get delayed by questions and answers until they acquire enough information to feel comfortable. However, if they operate from a macro perspective, don't waste time addressing the details. That kind of information will only bog them down in specifics they do not really want to know.

If you are a fire chief with a city or county manager, you may find your boss operates from either a political (council-led) or bureaucratic (manager-led) perspective. Your success will depend on understanding how their view of the world applies to decision-making in local government.

For instance, some managers will tend to operate from the perspective and belief that the political system is in place to provide direction to local government. Managers who operate from this perspective believe they insulate themselves from events by solving them in the political arena. These types of managers may view right and wrong only from the perspective of accomplishing what the council has directed.

Operating from a bureaucratic perspective, the manager may believe they are hired to provide direction to the council on best practices. From this perspective the manager will normally provide the council with solutions to problems along with the pros and cons of the potential options, ultimately recommending a best action.

You may find differences in styles are a real benefit. For example, if your boss does not like you and you don't like your boss, the two of you may decide to just leave each other alone. That kind of freedom can present you with some very real opportunities to be unique and creative in your job. Thus, while you may be frustrated over your relationship with the boss, your creativity and ingenuity could really shine.

The real key to working successfully with people who have an opposing style is to find a way their style can help you achieve common goals. Attacking their style or coming at them with a view of the world they do not want to see or understand will only confuse and frustrate both of you.

Work to the Middle of the Two Poles

If you find you are diametrically opposed to your boss's perspective concerning leadership, you should work hard to find some middle ground from which to operate. This can be explained as if you are at the North Pole and your boss is at the South Pole. Somewhere on the equator, there is a place both of you can stand and not be overcome by the weather, or in more work-related terms, where you can find some common ground.

This may be a little like the both of you leaving the table with a bad taste in your mouth; however, the taste is acceptable until you can brush your teeth. This type of conciliatory approach can ultimately be what saves your job or in the end makes it tolerable for your boss to work with you.

Get Out of the Pond and Watch the Turtles Operate

Leaders should take the time to climb up on the balcony and look down at the organization. The view from above is very different from the one at street level.

Regarding operating under the leadership of others, you may need to pull back and watch how your boss operates. In meetings, does your boss really listen and pay attention to the concerns of others? When there is a conflict, does your boss respect the opinion of others or do they argue one position and become resentful? Has a pattern developed that would indicate they hold a grudge when you disagree? Have your peers lost favor with the boss or been fired for reasons other than work performance?

If you answered yes to any of these questions, then you may want to look out. The one thing that will be certain is that if you see other ducklings being pulled under the water and you look or act like a duckling, you will most likely suffer the same fate. Remember you are learning, and the lessons learned are going to save you from the fate that awaits those not paying attention.

Focus On Something Other than Work

If you cannot leave when you find yourself stuck in a job or relationship that is not fulfilling your needs, it may be wise for you to step back from the situation and focus your energy on something other than work. I am not talking about complete withdrawal here, but rather concentrating on what brings you fulfillment.

No matter what diversion you choose, it may be beneficial to concentrate on something other than work and the fact your own needs and expectations are not being met. Perhaps a renewed commitment to your family, religious beliefs, or a hobby can provide a diversion from your current situation.

Remember, your work life should be just that: your work life. The money you get paid should be utilized to support some other important facet of your life. However, if your work is your life and you are too far gone to consider a diversion, you may want to start looking for another pond or recognize you are in your situation because *you* have not learned the lesson. Remember, you are responsible for *your* happiness.

Resist the Urge to Fall on the Sword

Most of us who assume leadership positions have a certain amount of competitive nature in us, and we can tend to view the world from a win-lose perspective. This kind of view sometimes translates as "I will show who is right and who is wrong." This usually leads to a bad outcome for you, while doing absolutely no harm to the other person.

First, you have a duty as a leader to support your organization and subordinates regardless of your boss's leadership. Finding you cannot support their position is one thing; however, deliberately sabotaging their efforts is wrong on just about every level of human behavior.

Second, and very closely associated with undermining your boss because of a disagreement, is the "take one for the team" approach. These are the people who will fall on the sword just to show their boss who was right. If you are considering this tactic, ask yourself how this demonstrates leadership in the minds of your employees. If you are indeed right about a situation and things are going to go bad because of your boss's leadership, then this is the time your organization needs you the most.

Falling on the sword is only going to result in your loss, thus satisfying no one. The world will continue to turn, fires will continue to be put out, and ultimately you will just be the guy who used to work here. Someone once told me that if you want to know how much you will be missed around here then place your hand in a bucket of water and pull it out. The hole that remains is how much you will be missed.

Seek Professional Help

Interpersonal relationships are important in all organizational environments. Understanding and working for leaders when we don't agree with their values or leadership style is something all of us will be confronted with at some point in our careers.

Before you elect to change jobs because of your boss's leadership, you need to really make sure the problem is not with you. If the problem is you, then switching jobs is not going to help. If you find yourself in a continual pattern of being unhappy with leadership, it may well be that you are the one who needs to change. For that you may need professional help. Being close to the situation may mean you need to find someone with a different perspective to help you see the cycle you are in. Often, simply breaking the cycle will create the learning needed to change the situation.

Remember the ducks that return to the pond each year only to have their ducklings fall prey to the turtles? The point here is if the same problems occur over and over, then just changing ponds is not going to help. Your behavior in the pond must change.

Move to Another Pond

When you have tried everything and your expectations do not match up to reality or when organizational decisions violate your value system, it may be a sign that it is time to move on. The important point here is you should not wait until the last minute to find a new pond. The time to change begins early, before you become so comfortable it becomes difficult to change.

I had a friend explain this to me using the analogy of two laddered buildings side by side with the bottom of the ladders touching each other. As you climb the ladder on one building, you'll notice you're getting farther away from the ladder on the other building. Soon, the ladder on the other building is out of your reach. If you decide you want to be in the other building, your only choice is to climb down and start from the bottom.

All of us have watched people climb too high on one ladder and then try to lean out to grab the other ladder. The result is usually a nasty fall. The lesson here is not to wait until you are so high on the ladder that it is difficult to make the climb down and start up on a new ladder. If you find you are too high on the ladder (you waited too long), you may want to just ride out the discomfort of your situation.

Do Not Burn Your Bridges

Make no mistake about it, the turtles in the pond talk to each other and in some strange way you need them. Do not acquire a bad name for yourself by placing blame for your situation on others. Remember, no matter the reason, you have no right to place the blame on anyone other than yourself. If you get fired, it is because you waited too long to switch ponds. If you are leaving because your style does not suit the style of your boss, then it is better for both of you. Remember, there are folks who will not agree with your leadership style as well.

Regardless of the situation you find yourself in, you need to resist the temptation of taking a parting shot. Remember, it really is a small world, and as my grandfather used to say, "If you don't have something nice to say, then keep your mouth closed."

Working with other people, even when we understand them, like them, and find they have complementary leadership styles, is not easy. When we do not understand them, like them, or find their leadership style appealing, it can be intolerable.

The important thing to remember is that your situation is determined by you. If you elect to work hard and your boss sees you are trying to find a middle ground from which to work, then you can usually survive. That said, if you are looking around and your fellow ducklings are disappearing, it may be time to get out of the pond or at least to start paying close attention to what they are doing wrong.

So, what is the moral of this story? There will always be snapping turtles at work, so running from their powerful jaws is not an option. Moving aside and letting them go for a weaker duck could be. If you find yourself a duckling in a pond full of snapping turtles, whether it is your boss, peers, or even subordinates, it is important to find a coping mechanism and stick with it until you are older, wiser, and more mature. Or, my personal favorite, find a mentor, someone you feel shares your values and can use their extensive wisdom and maturity to help you hone coping mechanisms that will last you a lifetime of wading in ponds with snapping turtles!

Part IV

Connecting Leadership: Making It All Work

High performance leaders connect the leadership dots by employing the abilities of all team members to solve organizational problems.

Maintaining Relevance

Ignoring reality is the first step to becoming irrelevant.

Key Points

- High performance leaders have a responsibility to prepare for change.
- Managing change maintains relevance.
- Understand relevance as it pertains to the fire service.
- Technology affects relevance when tools and techniques become outdated.
- Behaviors that contribute to irrelevance include complacency, stubbornness, and failure to know your business.
- Technology has increased the need for analytical versus anecdotal thinking.
- Understand value as it pertains to service delivery.
- There has been continuous progress since *America Burning* was published.
- Leaders should create a culture of innovation.

One aspect of senior experience is that you are provided a clear perspective on where things were and where they currently sit. I will stop short of saying that wisdom develops from the ability to look back because I am prone to making the same mistakes over and over if not careful.

Back in the day, fires were plentiful, and therefore we did not have a need to do anything else to maintain our perceived self-worth. However, the profession has changed and will continue to change. The way it was when we started and the way it will end up when we are done will look much different. For the older folks this has been a slow and often painful process to endure. For younger folks, the change has seemed more rapid but perhaps painful nonetheless.

Just like successful businesses in the private sector, we must continue to evaluate our business model and be responsive to changes in our environment. To ignore the changing landscape of our profession will undoubtedly lead us down a path where someone else determines our fate.

Fires will still need to be put out, and people will still need to be rescued. It will, however, be up to the more innovative and change-responsive fire departments to decide what they do between those events, which will determine how relevant and important they are to our customers in the future.

Early one day in March of 2021, my assistant asked me to meet with a citizen who wanted to come by and make a special presentation. I recognized his name because he was a descendant of one of our fallen firefighters from the late 1800s. He and his family were recognized when we dedicated the firefighter memorial, and he was also the manager of a tire shop just up the street from our headquarters.

Expecting a check for the fire foundation, or even cupcakes for the firefighters, I carved out some time to meet with him and show my appreciation regardless of what form the presentation took. As I watched from my office door, I saw him approaching the building with a trash bag that looked to be covering some sort of rectangular object.

After exchanging pleasantries, he said that his father had recently passed away and he had the dubious task of going through his belongings. As he untied the bag and carefully turned the picture into full view, I could see it was a photograph of a firefighter from days long ago. But like this was no ordinary picture, this was not just an ordinary firefighter. The firefighter in this picture was Joe Willard, the first recognized fallen firefighter from the Wilmington Fire Department (fig. 26–1).

Firefighter Willard was a member of Wilmington Hook and Ladder Company during the 1890s. He was killed in the early morning hours of June 17, 1893, from a wall and roof collapse at the Calder Brothers Warehouse on South 6th Street. Many firefighters were reported to have been injured in the collapse but none as severely as Firefighter Willard, who was pronounced dead at the scene.

I was very touched as the picture was presented because it dawned on me that it was 128 years old and, based on how old he looked in the photograph, was most likely taken just before he was killed. Even more remarkable, I came to find out later that the picture is the oldest recorded photograph of a firefighter killed in the line of duty in North Carolina.

For the next few days, Firefighter Willard was placed on a table in my office. One of my chief officers noticed that my recruit picture from 1978 was quite a

FIGURE 26–1. Joe Willard was born October 25, 1871, and died June 17, 1893, in the line of duty.

contrast when compared with the one of Firefighter Willard. He also took great pleasure in remarking that my recruit picture was quite a contrast to how I look today as well.

The contrast he described did get me thinking about how much change had occurred since that picture was taken of Firefighter Willard—and for that matter, how much change had taken place since I first joined the Virginia Beach Fire Department in 1978. I think it is safe to say that the pace of change has accelerated rapidly when comparing 1893 to 1978 to today. As a matter of fact, there may be more change occurring between any of our succeeding years than took place in the 85 years between Firefighter Willard's death and my being hired.

All this introspection got me thinking about change and how change needs to be managed to stay relevant.

Relevance is based on perceived value, and that is where things get tricky. The best way to describe relevance is to say that something is relevant to something else if it is closely connected or appropriate to the matter. So, if someone says to you that he likes fire trucks, and you say you have a son named Ryan, then those two things are not relevant to each other. However, if someone says they like fire trucks, and you say that you have a son named Ryan who is a firefighter who drives fire trucks, those things become relevant to each other. Shout-out to my son, Ryan, and my grandson, Mason, who happen to be firefighter medics who drive fire trucks.

Understanding relevance is important to the concept of how we change and adapt as technology, culture, and modern firefighting evolve. Simply put, the service we provide must be viewed as appropriate when compared with our customers' perceived needs. In addition, the value of the service must be viewed as appropriate when compared with the cost.

If you think that we do not understand this as a fire service, you need to look no further than our history. First, we were only responsible for suppression. Then came EMS, Hazmat, Tech Rescue, Community Paramedicine, and now Community Risk Reduction. We evolve based on need and to remain relevant to our customers. As the number of fires decreases, we add services so that our perceived value is greater than the cost associated with providing those services.

If you do not think relevance is important, then you have not paid much attention to history or how technology and essential needs change our customers' perspective on the value of a service. For instance, how many of you go to a library to research a paper? Hardly anybody. So, if you are a librarian by trade, how much longevity do you have? Flown on a plane lately? How many flight attendants did you see? For that matter, whatever happened to the Walkman, paper maps, Polaroid cameras, pay phones, and—for all you instructors out there—overhead or slide projectors?

Times change. Some tools and techniques become irrelevant, and newer and improved technology becomes more relevant. The key is to understand your business and what your customers expect from you and then to find a way to balance the cost associated with that service as it is compared to your customers perception of its value.

Kodak dominated the photo film market during the 20th century. Kodak leaders during the time failed to see digital photography as a disruptive technology. As a matter of fact, a Kodak engineer invented the first digital camera; however, company executives, ever committed to film, remained reluctant to approve this new technology. The company thought they were in the film business and did not understand that they were instead in the picture business. Kodak filed for bankruptcy in 2012.

As another example, Nokia was a very successful Finland-based company and the global leader in cell phone technology during the 1990s and 2000s. When the internet came along and data, not voice communications, ruled the day, they failed to adapt. Rather than focus on how data would affect their brand, they continued to focus on hardware and not software. The final nail in their coffin came when the iPhone launched in 2007 and it did not have a keyboard. Once again, we find a company that thought they were in the cell phone business, and they were actually in the communications business.

Then there was Xerox, a company that was the first to invent the personal computer, but thought that digital technology was too expensive. Just like Nokia, Xerox failed to see digital products as a threat and instead assumed the future was in copy machines.

The last example is Blockbuster, a video company that based its business model on fixed locations providing VHS tapes for rent. They were at their peak in 2004 and survived the VHS to DVD technology change but maintained their delivery system fixed-location focus. Then came along a company called Netflix that started shipping DVDs to people's homes. Netflix, understanding they were in the entertainment industry and not just the DVD business, continued to leverage technology and morphed from DVDs to a streaming service that is now a multibillion-dollar company. Blockbuster on the other hand went bankrupt in 2010. As a side note to this story, Netflix proposed a partnership with Blockbuster in 2000 but was turned down.

Consider how these possible causes for resistance to innovation might parallel leadership thinking that is prevalent in your organization:

- **Complacency:** This can be described as "Things are going well, and hardly anyone is rebelling right now, so let's let sleeping dogs lie."
- **Motivation:** "We are making good money right now, so why put forth additional effort? Change is hard, and I would rather just sit here on the porch."
- **Courage:** Perhaps lack thereof. It takes leadership courage to navigate change and keep your organization and its business model relevant.
- **Value:** Leaders must understand how their product's value, as perceived by the customer, determines future success.
- **Weak Signals:** Subtle changes in your environment can lead to big changes in your business.
- **Technology:** Failure to match new technology to your product's purpose and function may jeopardize your organization.
- **Group Think:** Passive leaders believe the notion that when we all agree with each other, everything will be okay.
- **Modification:** Organizations must constantly reinvent when presented with challenges.
- **Stubbornness:** Refusal to accept or recognize changes that will affect the value of your product means you will be left behind as times change.
- **Knowing Your Business:** As a leader, you must understand the true nature of your business and what it is your customer needs from you.

We could probably use each of these tripping points as current examples of failures in our own organizations. That said, everyone should take a deep breath as the sky is not falling. Fires and fire departments are not going away anytime soon. You and I both know that accidents will happen, and dumb people will make mistakes as they roam the streets of our communities. Besides, in every survey ever taken, our customers love us, respect us, and appreciate us for what we do.

Despite this appreciation, our bosses and our customers demand to know the value of what it is they are getting for the money they are paying. That will not change anytime soon. If you are in an urban setting and run lots of fires, you will be seen as relevant and necessary. On the other hand, if you have a department full of firefighters who are sitting around the station and not fighting fires, you will have a different dilemma.

Understanding types of value will help when we consider how we might adjust our services to remain relevant to our customers.

- **Functional Value:** You provide a service or product that is needed or hard to find.
- **Monetary Value:** You save or make money for your customers (think insurance).
- **Social Value:** Your service or product is necessary for society to function.
- **Psychological Value:** Your service or product is needed to make people feel safe.

What makes value interesting to consider is that the sources of value—and the degree to which each is important—vary from situation to situation and from customer to customer. In this regard, value is not equally important in every circumstance.

To further understand value, you need to have a grasp on the term "value proposition." A value proposition is a cost-benefit formula that is evaluated subconsciously and automatically in your customer's mind when they encounter your marketing touchpoint. This equation compares the perceived benefit and actual cost of transacting your brand or product.

Think of the example of buying an ice cream cone. When you decide you want an ice cream cone, you subconsciously have in mind what the value of that purchase will be compared to its actual cost. You expect to pay a little more for two or even three scoops as you determine that the value is greater even if the cost is higher. The question remains, just how much more will you pay?

In this example, I suspect you would be okay if the one scoop was $3.00 and then each additional scoop was $.75 extra. You are thinking to yourself that you can have one scoop for three bucks or for less than five you can get three. What a deal, huh?

But what if they served up your three scoops and said it would be $17.00? I will bet that you are just like me and would say you are not paying $17.00 for a three-scoop ice cream cone. Your faculties would suggest that this is not a good deal and the value of the ice cream is not worth the money you are having to pay.

There are two types of value propositions, and the difference is whether you are a for-profit or not-for-profit business. In a for-profit business, the value proposition is based on an equation that includes your product, its features, the number of customers that will purchase the product, and then the net profit produced by those sales. In a not-for-profit or service delivery organization, the service elements, cost of service, and number of customers still applies; however, the value proposition is almost entirely based on perceived value.

This leads me to gold plating your value. Gold plating, as was described in an earlier chapter, refers to offering a product that is very complex with high costs and will appeal to only a small segment of your customer base. Remember the DVD player? On the one hand you could purchase a standard player with only a few options that is a good value for the money. Contrast that with a high-end player that has many options and will only appeal to a small segment of your customer base. The good value player will appeal to most of your customers while the more expensive and complex player will not. In this example, if you concentrated your marketing efforts on selling the high-end player, you may make a good profit on each sale, but your volume of sales and overall gross profit would suffer.

How does gold plating apply to the fire department? Consider the many types of programs you could offer your customers based on the demographic characteristics of your community. Say you are protecting a small town, but as the fire chief, you insist on having an Urban Search & Rescue Team. In that same small town, you have only one building that is four stories tall but insist on having a ladder truck. Perhaps you also insist on having a fully paid staff when the company only makes 10 runs a month.

The purpose of this discussion is to help you see that the fire business and our success is mostly based on our customers' perception of our value as it compares to the cost. At some point, just like the example with the ice cream cone, your value compared to cost will be a factor in determining the success of your organization and the willingness of your customers to continue to fund your efforts.

This begs the question, what is our customers' perception regarding value? Well, in the first place we are heroes in the eyes of our citizens. Because of our nature to be humble and unassuming, everyone likes us, and we are seen as honorable and dependable. In general, our facilities are specific to the community and seen as community amenities. In other words, the fire stations are neighborhood places that connect people to a common identity. Maybe more important is that we are viewed as a necessity, and while our costs are rising, they are still viewed as tolerable when faced with the alternative of not having them.

What is it that has changed or will be changing about our profession? To answer this question, look no farther back than 1973. In that year, a book called *America Burning* was published by the National Commission on Fire Prevention. The book was commissioned because, guess what, America was burning. Fires "back in the day" were frequent and larger in scope, and when they occurred many people were killed and injured.

The final *America Burning* report focused the fire service on three key objectives: Address the built environment through building codes and code enforcement, enhance education and prevention efforts, and improve firefighter training. The report concluded that fire prevention and public fire safety education were critical to reducing the losses associated with fires and that firefighters needed to be better educated for their jobs.

The results of *America Burning* dictated buildings be constructed to a stricter code that ensured they were resistant to fire and fire spread. Many new buildings are now either sprinklered or outfitted with early detection devices so fires that do occur are smaller and more likely to be held to the room of origin.

The results of *America Burning* dictated buildings be constructed to a stricter code that ensured they were resistant to fire and fire spread. Many new buildings are now either sprinklered or outfitted with early detection devices so fires that do occur are smaller and more likely to be held to the room of origin. Education is also a focus through both early childhood and adult education efforts, as we can reach almost all people through one type of media or another. Prevention of fires through code enforcement is now almost uniformly standardized around the nation. Lastly, firefighters are better trained through certification requirements than at any previous time in our profession.

Buildings are newer and safer; the public is better educated in fire prevention; despite an increase in population in the United States, overall fires are down; fire deaths per capita are down; fires are less frequent; fires are much smaller and more likely than not to be contained to the structure of origin; and through the development of the career fire service, more firefighters are now on duty and closer to places where fires are more likely to occur.

Just in the small window of my fire service career, I have born witness to a completely different fire service than the one in which I started. As an example, in Wilmington, fires are down 1.41%, fire deaths are down 8.33%, and injuries are down 43.75%, and all of that is over one decade and despite a 17% increase in population. Even more amazing are the statistics related to fire containment. Fires that do occur are held to the room of origin in a greater percentage of cases than during the previous 20 years. This is a great trend and one that I believe will continue as redevelopment occurs.

These trends can be directly attributed to a few factors. The first is that Wilmington can no longer annex unincorporated areas of the county, as the law in North Carolina was changed to restrict forced annexation. The effect this had on property values was significant, as the amount of available land is now finite. This caused the land to increase in value, and redevelopment subsequently occurred. That development is being built to a different code.

While the resulting change in density does correlate to some increase in calls for service, those calls are more specific to EMS and not fire. The other unfortunate effect is that redevelopment is causing gentrification as many lower income citizens cannot afford to live in the newer, more expensive developments. As is the case all over the world, much of our fire problem is directly proportional to our customers' socioeconomic status. In other words, while we are becoming denser, we are becoming denser with customers who make more money and require less service from our fire department.

Each of our situations is different, and I can only provide a few observations about Wilmington. You can then use your judgment to determine if these same situations apply to your department. That said, I will go out on a limb and predict that they do, and if you are not considering them, you will need to in the future.

Another fascinating movement in modern and progressive fire departments is that they are now looking at their environment through a risk reduction lens. Instead of just running fire and EMS calls, they are assessing the totality of risk across their boundaries and deploying strategies to reduce the effect of those risks, using risk reduction thinking to drive change and innovation. This constitutes more of a way to think and behave than it does a single process to follow.

The key to the risk reduction methodology is that it forces you to look at analytics to determine customer need. It steers your organization to address areas that are clearly in need of attention and in doing so keeps you relevant and a good value when evaluated by your customer.

Let us not kid ourselves and say that there are not challenges to being innovative and maintaining relevance. Institutions like the Insurance Services Office (ISO) require us to behave in a certain manner even if it is not the best fit for our communities. Additionally, most of our frontline folks are close to the delivery aspect of what we do and have a hard time seeing the big picture. History, culture, tradition, pride, and honor all play a role in making change hard to implement even when it is necessary.

What are the expectations of leadership to make certain your organization is relevant and a good value for your customer? Well, the first is that it is your job to prepare your organization for the future. That is why they call it leading. Your responsibility is to match your organization's efforts with the actual needs of your customers. This sometimes requires you to lead people to a place they will resist or may see no real value in reaching.

The process will be much less stressful on you and your employees if you have established a culture that is accepting of change and therefore quick to adapt to a changing environment. That can happen by default if you are committed to hiring people who are well suited for both what is required now and what may be required in the future.

Creating a culture of innovation to maintain relevance is no easy task, as it requires one to focus efforts on collaboration and teamwork. Collaboration and teamwork can be achieved by creating clear channels of communication that go both up and back down the organizational structure. In addition, you should recognize people for thinking outside the box and for using technology to create efficiencies in operations.

> What are the expectations of leadership to make certain your organization is relevant and a good value for your customer? Well, the first is that it is your job as a leader to prepare your organization for the future. That is why they call it leading.

Placing the right people with the right skills in the right place at the right time is of the utmost importance. You should fight the urge to place people in positions because of rank and instead match their strengths to what needs to be done for the organization to be successful.

As a leader, you should also do everything you can to diversify your organization. Diversity of thoughts, opinions, and experiences are the foundational blocks on which progress is built. It will also help you as a leader make certain that proper and thorough consideration is being given to understanding your customer's needs.

This entire discussion on maintaining relevance boils down to how leaders guide the organization through phases of innovation, basically, having the courage to question everything and not just let perception be the basis of reality. Challenge the way you think and the way other people think, and have discussions with your folks that are difficult and uncomfortable. Whatever you do, never assume that things will just stay the same. Things never stay the same for long.

The most important task in remaining relevant is understanding your business to the point that subtle changes in your perceived value can be recognized quickly. We mentioned earlier how being able to see weak signals while they are weak is vital in today's fast-paced world. That very faint light you see coming at you could very well be the headlights on an FA-18 if you are not careful. Not being able to see how subtle changes in your environment can lead to big changes in your business will quickly lead to irrelevance in the minds of your customers.

This could be as simple as employing new technology to change one aspect of your process chain to newer and more effective products that are appealing to your existing customers. This requires understanding how your product's value, as perceived by your customer, determines your future success, avoiding failure to match new technology to your product's purpose and function.

This is a huge challenge and can be very difficult for those of us in public government because in most cases we are the sole provider of our services. Additionally, we don't have a value proposition focused on profit, which is very carefully monitored by for-profit organizations.

All of us from time to time lose motivation. Perhaps we are at the end of our careers and just don't have the same enthusiasm for pushing things uphill all day. Or we may just be plain stubborn and refuse to change even when we are fully aware that modifications in our practices and procedures are necessary.

Maintaining relevance will not be hard for you or the organization if you allow the data to speak to you and you react to follow trends in the data. As a leader, you should always strive to understand what efficiency looks like in your operations and have the courage to lead your organization toward those efficiencies. Whatever you do, never lose your courage to lead. If you do, please consider stepping aside, as the leadership game is hard work, and your people deserve someone who will work hard.

> I am reminded on a frequent basis that no journey ends—rather another just begins. Perhaps a better way to describe this ever-changing part of our human experience is to as a road with many turns, hills, and valleys, each one just another part of the journey we call life.

I am reminded on a frequent basis that no journey ends—rather another just begins. Perhaps a better way to imagine this ever-changing part of our human experience is as a road with many turns, hills, and valleys, each one just another part of the journey we call life.

If relevance is what we seek, then it is irrelevance we must shun. That may be uncomfortable for some, but staying on the same road because you are unsure what may be around the left or right turn cheats you out of the thrill of a new and potentially powerful experience. After all, new experiences are what make life interesting and compelling.

Just as in your personal life, your organization will be faced with many hills, valleys, and turns. Do not be afraid of what might be around the next corner.

Enjoy and embrace the journey of constant improvement and open communications with your customers because that is what is necessary for you and your organization to maintain relevance, now and in the future.

For those of us who do not like change, I say hold on to your hats. The winds of change are blowing harder every passing day.

High performance leaders understand their responsibility to prepare for and manage change. For a leader, the ability to manage change is a trait worth its weight in gold in these fast-changing times.

The fire business has been changing since the late 1970s, due in large part to the implementation of strategies recommended in the *America Burning* publication. The reduced number of fires has freed up capacity for urban and suburban fire departments to get involved in EMS, Hazmat, Tech Rescue, and now Community Risk Reduction.

As we continue to see a decline in fires, there will need to be a concerted effort to think about how we maintain relevance. This does not refer to relevance in the sense that we are somehow not going to be needed anymore, but rather in how our value to the community is judged based on the ever-increasing cost of fire protection.

In the past, city managers and budget planners relied on fire chiefs to support budget requests, and for the most part they were successful getting what they needed to get their jobs done. These days the technology that modern fire departments use to analyze everything from response times to reliability factors provides data that is available to budget planners and staff. This has changed what we do from anecdotal to analytical.

The good news is that there are specific behaviors that contribute to irrelevance. Being on the lookout for these organizational behaviors is a step in the right direction to create a culture of innovation.

A View from the Clouds

*Son, remember, you are always only three bad
decisions in a row from being at the bottom.*

—Cecil V. Martinette Sr.

Key Points

- It is easy for leaders to lose perspective on the needs of their employees.
- Relationships are created at ground level while speaking to and caring about people.
- All of us are only one bad decision or unfortunate event from a disaster.

If you are a chief officer reading this book, you likely realize that some days, it is easy to get separated from your workforce. The fast-paced nature of modern organizations has the top leaders running from this meeting to the next, only occasionally stopping to talk with an assistant for the purpose of organizing and scheduling even more meetings. As an example, I am sure you have heard the old "We never see the chief" line from your subordinates. If you are on the line, I am sure you have said it one day or the other. I know this because I was once one of the ones saying, "We never see the chief."

At this point in my life, I have become comfortable knowing that there is a paradox to all this notion that "the chief never comes by." Even though it is not unusual for this situation to occur, truth be known some of these same folks would be describing you as a micromanager if you spent all your time out in the field, not to mention how ineffective you would be with getting your own set of responsibilities accomplished.

I think what we are dealing with in these situations is a loss of perspective on the part of the leader and the followers. As time goes by, we tend to forget the challenges each of us faces in wrestling with the ins and outs of the daily grind. We lose an appreciation for each other that comes from a perspective that is narrowly focused with no room for consideration of challenges other than our own. This is both an "us" and "them" problem, and both "us" and "them" need to work on an appreciation for each other.

In my career, I have flown in a helicopter above many disasters. I have found that it is easy to lose my perspective while flying high above these situations. Sure, things look bad, buildings are collapsed, streets are flooded, and people are obviously in peril. Still somehow, I can become disconnected from all that personal tragedy when I am in an H-60 Blackhawk.

I would describe many of these flights as surreal. As you view the landscape, you are disconnected from the millions of people whose lives are now vastly different than they were just the day before. Disconnected from the husbands, fathers, wives, and mothers who perhaps find themselves hundreds of miles from home and unable to assure their children and loved ones that things will be okay soon, unable to assure them they have a home or even where they will get their next meal.

These were not my observations from the air. Rather, these were the people I saw while on the ground, some of them walking aimlessly with nowhere to go and nothing to do and still others waiting in lines or standing in hotel parking lots and grocery stores for no other reason than what else was there to do?

On one of my disaster deployments with the Federal Emergency Management Agency (FEMA), I served on the Rapid Needs Assessment (RNA) Team. This multifunctional team works for the Regional Office of FEMA to assess overall disaster damage and then recommend the placement of assets to help speed along mitigation and recovery efforts. On this deployment, our team had representatives from water control, spills and leaks, electrical services, transportation, mass care, and me as the Urban Search & Rescue (US&R) representative (fig. 27–1).

In this scenario I am not describing only those types of people who live on society's edge (although of course they are included) but rather men and women with children from well-kept neighborhoods, with PTA meetings and soccer practice schedules who suddenly find they have been dealt a crushing blow to their very existence.

Now, I am no stranger to hurricane deployments with FEMA and the many tasks required to accomplish just our part of the recovery process. This one, for Hurricane Rita, had me using both hands and at least one foot to count them all. Normally I am an incident support team operations or branch officer for the US&R

FIGURE 27–1. The author serving for FEMA during an Urban Search & Rescue Deployment (courtesy of Jocelyn Augustino).

response. I am comfortable on the ground at the disaster site. That is where the rubber meets the road in rescuing people and is the function I get the most satisfaction from as a participant.

On this deployment, my focus was on collapsed structures in the disaster area and the number of people who could be potentially trapped in buildings based on the type of construction. The idea was to assess damage very quickly and then recommend the placement of US&R teams based on the number and density of collapsed structures in any given quadrant.

The RNA team tries to get up in the air as quickly as possible after the storm has passed. For the people on the ground affected by the disaster, I am certain there is a different definition of "quick." That said, we must fly in helicopters, and as my friend Chase Sargent use to say, "Helicopters don't really fly; they just cause so much commotion the earth backs away from them."

As the first flight began, I had no idea what to expect. Like I said before, I am usually a ground guy on these scenes. Briefed by the crew and ready for flight, the weather was just breaking, but it was still very windy with periods of rain. Getting tossed from side to side in a helicopter is no easy feeling for most of us flatland folks. However, I eased my fears by keeping in mind that the pilots didn't want to crash any more than I did. In that respect, I had a lot in common with the pilots.

As the H-60 started its rise in altitude, I got my first opportunity to view the landscape as a snapshot picture of our society's many parts. It was then that it hit me that we live in a very complex, interdependent myriad of systems that must operate with some degree of reliability for us to function effectively as a society. If even one of these systems goes out for some reason, our way of life changes very quickly.

Just consider for a moment how dependent we are on water for our very existence. Stop the water, and guess what happens? Suddenly dehydration becomes an issue, and then other health issues arise without the ability to flush waste products to a sewage system or even bathe to maintain control of germs and disease. And all of this can occur in mere hours, not days or weeks.

From my window, I could see electric towers with their many poles and lines, water plants, sewage treatment plants, interstate roads, businesses, malls, churches, single and multifamily housing, hospitals, and many more types of buildings and other infrastructure that we find ourselves in on any given day. Think about how many of these types of buildings you pass each day on roadways you depend on to make it from point A to point B. We do indeed live in a complex world.

The next thing I noticed from my new window to the world was that at that point nothing was wrong with any of it. Cars were moving through an organized network of roads, and everything was geometrically situated and orderly. Our flight took off as soon as possible after the storm abated, originating from an area generally unaffected by the storm. However, traveling at a high rate of speed in a straight line, it did not take very long to go from orderly to dysfunctional.

The first sign that anything was amiss was the trees: tall trees bent in the direction of the strongest winds, shorter and less substantial trees with their tops broken off and limbs hanging precariously in one direction or another. *No big deal*, I thought to myself, as these types of things happen in routine thunderstorms and downbursts.

Debris was one of the second things I noticed: in the roads and stuck in trees. Paper and other signage were scattered around or piled up against the backdrop of buildings and fences. Food wrappers and containers of every type littered the landscape. The further we traveled, the more different types and sizes of debris we noticed, first lightweight materials and then heavier, more substantial building materials.

As the flight moved forward, things changed very quickly. Shingles were missing from roofs, and the occasional piece of siding was laying in the yard of a single-family home. A few more trees down, some of them were leaning on the very houses where they once provided some relief from the sun.

In mere minutes, the scene became ghastly: No movement below. Only missing roofs and buildings twisted and collapsed under the strain of high winds that pounded them from every direction for hours. Vast stretches of land with trees snapped off at their tops. Trees lying across roads and downed power poles. Cars resting upside down, obviously not in the same place their owners left them. Construction materials and people's personal effects littering the streets of the neighborhoods they once called home.

Other areas were completely devastated by tornadoes, taking the appearance that land was carved with a chisel between other only moderately damaged properties. The waterways below were littered with boats, upside down and run aground on shores without docks. Oil and other fuels could be seen as slicks on the water for miles. Still, I thought to myself, *This could have been much worse.*

It was difficult to step back from the intimate nature of the damage at that point because the flight had remained a very impersonal experience. When looking at large collapsed metal electricity towers, flooded water plants, and nonoperational sewage treatment plants, it somehow became almost more than the mind could comprehend. After all, while we depend on these utilities, they are not the essence of life and certainly not personal.

As we flew from one grid section to the next, I quizzed the pilots on our location. I also could not help but hear them on the radio referring to the local places from a perspective much different than mine. At first it was something like, "Wow, look at the damage to that Target store." Then it became even more personal as they looked at the damaged and destroyed homes of friends and acquaintances.

That is when my view of the situation took a different perspective. From 1,200 ft. the situation wasn't about how widespread the damage was but rather how Joe's house wasn't there anymore. And I just happened to be with someone who knew Joe and cared for him and his family. I could hear the despair in the pilot's voice as he hoped and prayed that Joe and his family made it to safety before the storm hit.

For a moment I was struck by the personal nature of this tragedy. While I was concentrating on the job at hand from a more global perspective, I could not help but think that each one of those homes and businesses belonged to someone. The churches below without roofs were at one time places of worship for members of the communities. All of this caused me to reflect on my family and home and how lucky I was to be flying above the disaster and not a part of it.

To put a disaster in perspective, consider that most of us do not handle the situation very well when the electricity goes out. Now think about a situation when everything is out, and you don't even have a way to escape. In a moment of poor decision-making, you find yourself going from a fully functioning part of society to waving from the roof of your house in hopes that a helicopter will take you away from your misery.

As the flight continued from one critical part of the grid to another, the same situation played out. From the outer edges of the grid area to the center of the devastation, the scene changed from moderate to significant: businesses and homes from damaged to destroyed. The magnitude and sheer miles of damaged buildings were almost overwhelming.

As our flight made its way back to the Air Force base, the weather had improved, and we once again arrived at an area mostly unaffected by the storm. At that point all of us on the flight were mentally and physically exhausted and still facing hours of reports that needed to be completed back at the joint command post.

Leaving the base that day I remembered that I did not pack any running shoes, and the thought of spending 10 days in my work boots was not something I was looking forward to. Contemplating the situation on my ride back to the command center, I stopped at a Walmart to pick up a cheap pair of hiking shoes. As I entered the parking lot, I was struck by the number of people milling around. Upon further scrutiny, I noticed people were taking refuge from the heat under trees, and many of them looked like they were camping. Kids were throwing footballs and families were sitting in small groups, appearing to be just passing time. *Odd*, I thought at this point but what the heck I was not from the area and perhaps that was what people did here.

When I got to the checkout counter, I asked the cashier what all the people outside were doing. She told me it was not unusual to have people outside the store at all of hours of the day and night. She further informed me that the shelters were full, and these people had nowhere to go. They ended up in front of the store because the store has food, clothing, restrooms, and other life essentials, and it is open 24 hours a day.

On the way back to my car, my perspective of this tragedy had changed dramatically. These people had faces filled with uncertainty and despair. Mothers with infants, children, grandparents, people in wheelchairs: all I am sure hoping their nightmare would end soon. At the very least I am sure they were hoping that nightfall would bring relief from the heat and sun.

Imagine for a moment having to camp in the Walmart parking lot just to make sure your family has a place to go to the bathroom. Many of these people looked hurt and tired, and I was filled with anxiety over their situation. As I left, I felt somewhat guilty for getting back in my air-conditioned rental car for the ride back to the command center.

Later that night I had the occasion to meet a man as I was walking to my hotel room, which was located miles from an affected area. As I passed the pool, I realized I had left my notebook in the car and turned around to head back to the parking lot. The 40-something-year-old man left his table and walked toward me as I approached my car. He inquired as to whether I was leaving, and I replied no, that I had just forgotten something.

As both of us turned and walked back toward the pool area, he told me that he had nowhere to stay and was hoping I was checking out of the hotel. He stopped here because he was out of gas and, as luck would have it, the place had a pool. At least the kids would have something to do, and all the swimming would make the night's stay in the car a little easier for them, he said. I was saddened and once again reminded that the people caught up in this situation were not unlike me.

I cannot help but wonder what twist of fate separates me from some of these victims and how close all of us could be on any given day to becoming one of them. How fragile our very existence is, and how quickly can all of that change. From happy to sad, from prosperity to destitution: it can all change in the blink of an eye.

Over the course of Hurricanes Katrina and Rita, I was witness to some of the very best and worst a society can offer. People standing in endless lines not even sure what they were in line for, just knowing that whatever was at the end was more than they had now. Dead bodies on the side of the road, and people going to the bathroom in bushes because that was all they could do. All this amid thousands of folks like me just trying to do their part to make things a little better for these unfortunate people.

In my life I have always been able to disassociate myself from the personal nature of disasters. People dead in cars, killed in fires, crushed in collapsed structures, or drowned in floods were all people who for whatever reason ended up in the wrong place at the wrong time. On occasion, I would even make sense of the situation by thinking that in some cases the victims shared responsibility for their fates. We all have our ways of dealing with situations that make us uncomfortable or that we cannot seem to make sense of at the time. I feel terrible about that now.

During the next day's flight, the magnitude of personal tragedy took on new meaning. Each home down below seemed to belong to one of those families I met the day before. Each open room was the room of one of those kids swimming in the hotel pool or camping in the hot sun outside the Walmart. It was truly sad, I reflected as I gazed out my window to the world.

My observation here is that humanity cannot be viewed from the air; it can only be felt on the ground when in personal contact with others. It is easy for those of us in the rescue business to become disassociated from the tragedies to which we so willingly respond. Our participation in these events should not be taken for granted as many of us work to bring order back to society, hundreds and in some cases thousands of professionals, some risking their own lives for people they do not even know. We will always need people like that.

I can now say that our ability to successfully rebound from these events is not necessarily determined by the response itself, but rather by how each of us reaches out to help our fellow neighbor, each of us realizing we are only one bad decision or unfortunate event away from being in the same place as the people we see on television: standing in line or sleeping in the Walmart parking lot. In situations like these, the government is not going to save us, and likewise no rescue team is going to make things better. Only individuals in our midst who are willing to take personal responsibility for helping each other will make these occurrences less tragic.

Final Thoughts

I usually come home from these events filled with stories about rescues and the brave men and women who carried them out. Lives saved and those unfortunately lost seem to always make up these stories, as do the lessons learned that can be applied to future rescue situations in hopes of being more efficient and effective. All these are powerful lessons that can be applied to my life and those of others who work with me.

With respect to lessons, this disaster was no different. However, sometimes these lessons come about in situations where you would least expect them, and perhaps these are life's most powerful teachings.

As the leader of any organization, your responsibility to your team is great. This responsibility never goes away and is indifferent to whether they even like you. And believe me when I tell you that some of them will not like you.

The important lesson is that you cannot truly understand the perils and troubles of your people by staying in your office or maintaining the 30,000 ft. view. It needs to take place at the kitchen table where you hear them speak and can sense their frustrations.

As you move forward in your career, always view humanity from the ground. Value each day as a gift and in some form or fashion give this gift to others. Forgive people. Never give up on people. Love people.

28

Final Thoughts on Leadership

Leadership is many times about taking people to a place they can see no value in going and therefore find no specific urgency in getting there.

I have fielded many questions over the years from fire department officers about leadership sustainability. Rookie officers are often looking for a specific set of rules that will assure their future success. Veteran officers who have fallen from grace are seeking insight as to where they went wrong.

No one set of rules, values, or commitment to self-discipline assures success in the fire service. That's why leadership books remain some of the most sought-after reading material. Everyone wants to try and figure out the leadership holy grail, and in my view, it just is not there to be found.

For many years now, I have kept a journal that I call "My Thoughts." Many of the stories and chapters in this book had their beginnings in the rambling notes written in this journal. One of the folders in the journal was a collection of random thoughts concerning situations in which I have experienced something that I wanted to remember or maybe would find useful in a future situation. Usually, these thoughts were lessons learned either by my own first-hand experience or by observing the positive or negative leadership traits of others.

Many of these thoughts end up on a page I call "My Rules for Leading." They have helped me over the years keep my focus on what is important in leadership, and I thought I would share them with all of you.

- **Build relationships:** True leadership rarely takes place in a group setting. Rather, it is done one person at a time, and success is almost always proportional to your relationship with that individual.

- **Make others powerful:** Instead of accumulating your own power, try to give your power away. Effective organizations are run by people of all ranks who have the power to get the job done.
- **Take the light off yourself and place it on others:** Do everything in your power to deflect notoriety. In most cases, it is others who have made you look good, and to take credit for that is selfish.
- **Build capacity in others:** The more capacity you build in those who surround you, the greater your chance of leading them somewhere.
- **Practice your profession like it is art:** The result of our work is painted on the minds of our customers in the same lasting way that a painting leaves the artist a legacy. Make sure you paint well each time you get to paint.
- **Have strong personal values:** Personal values drive how you behave professionally. Figure out your personal values and never, ever violate them.
- **Have faith:** Believe in something, but most definitely believe in people and their willingness to do a good job.
- **Be humble:** You are never top dog. No matter how high you climb or how important you think you are, there are others who are higher and more important. You have faults and you will make mistakes, so be sure your support system is willing to support you.
- **Do not get mad first:** Try and understand a situation and learn everything you can about it before deciding to be disappointed.
- **Focus on the good aspects of everything:** All of us make a choice in life to view things in either a positive or a negative fashion. Some good can be found in every situation.
- **Embrace change:** No need to add anything here. Refer to the chapter on relevance.
- **Take care of the inside, and the outside will take care of itself:** If everything you do is completed in a manner that considers the impact on the inside of your body, then the outside will be exactly like it is supposed to be.
- **Have the courage to take the next step:** Do not get caught up in trying to please everyone or trying to find approval at every turn. If your job is to lead, then you should lead, even though that takes courage.
- **Leave a legacy:** Make sure that when your chapter is written someone has something good to say about you and the impact of your life on others.

Final Thoughts

In the previous pages we have talked about all these rules for leading; however, know that you can do these things and not be successful all the time. Leadership is tricky. It is not necessarily made up of knowing the things to do but rather understanding when those things need to be done.

Some of the world's most respected and seasoned leaders are likely to wake up one day and find themselves on the wrong end of a front-page newspaper story. The higher you go up the leadership chain, the greater the external expectations become for how you conduct yourself both personally and professionally.

> Leadership is not necessarily made up of knowing the things to do but rather understanding when those things need to be done.

If for some reason you do find yourself on the undesirable end of fate, I hope someone who has learned life's most important lessons reaches out to you and lends a helping hand. In the meantime, perhaps it is time for you to reach out to others. Now go lead.

A

Organizational Assessment

Exercise

To establish a baseline for your organization's efforts at creating a high performance workforce, use the following instrument. To create diversity of thought, make certain that you and many others who are at various levels of the organization complete the survey. Key takeaways will be where your organization stands regarding each of these principles and whether there is agreement between you and the various other key stakeholders. Even more analysis could be explored by asking whether there is congruency between organizational levels or whether some of the organization's levels differ in opinion.

To what degree do we as an organization:

Utilize consistent, published, and defined values in order to promote consistent organizational decision-making .

Always Almost Always Most of the Time Rarely Never

Develop and nurture quality personal traits that can be patterned over time in order to demonstrate consistency in leadership behavior.

Always Almost Always Most of the Time Rarely Never

Coach, mentor, and lead by example in order to build relationships and promote mutual respect.

Always Almost Always Most of the Time Rarely Never

Provide leadership opportunities in order to encourage career development and succession planning.

Always Almost Always Most of the Time Rarely Never

Draw upon the insights of stakeholders in order to make well-informed decisions.

Always Almost Always Most of the Time Rarely Never

Foster a culture of continuous improvement in order to be adaptive and responsive to a changing environment.

Always Almost Always Most of the Time Rarely Never

Encourage creative thought in order to promote innovative and effective solutions.

Always Almost Always Most of the Time Rarely Never

Promote teamwork in order to create a collaborative work environment.

Always Almost Always Most of the Time Rarely Never

Acknowledge the good work of employees in order to show appreciation for their value and contribution to the services we provide.

Always Almost Always Most of the Time Rarely Never

Hold each other accountable in order to assure personal ownership in organizational performance.

Always Almost Always Most of the Time Rarely Never

Use diverse, open, and interactive communication methods in order to promote employee engagement.

Always Almost Always Most of the Time Rarely Never

B

Traits Assessment

Trait	Always good	Occasionally good	Neutral	Occasionally poor	Always poor
Humility					
Credibility					
Trust					
Being nice					
Fairness					
Respect					
Responsibility					
Vision					
Communication					
Optimism					
Patience					
Situational awareness					
Evaluating expertise					
Confidence					
Taking credit					
Charisma					
Authority and control					
Humor					
Delegation					
Happiness					
Hope					
Critical listening					
Adaptability					
Self-discipline					
Courage					
Damage control					
Mentorship					
Followship					

C

Planning Team Charter

Team Mission

Interrelationships

The Strategic Planning Team will use a strategic planning process (beginning with the development of this charter) to identify stakeholders and gather input by providing a platform of open communication throughout the organization while striving to align with the vision set forth by City Council, the Strategic Leadership Team, and our customers to provide quality, professional services to the community and a rewarding workplace environment for our members.

Roles

All Team Members

- Have proper and punctual attendance
- Display positive attitudes (observable and measurable)
- Have commitment
- Actively participate
- Lead by example (walk the talk)
- Have full team membership
- Help perform work
- Participate
- Engage
- Ask questions
- Give encouragement and show curiosity

Chief

- Serve as team leader
- Assure team accountability
- Keep all pieces together

Senior Staff, Battalion Chiefs, Captains, and Firefighter/EMT/P

- Interact with peers to give and receive input
- Act as a conduit between team and rest of staff
- Educate
- Perform drills
- Distribute and run specific, related drills
- Gather, analyze, and compile data
- Provide feedback into process

Facilitator

- Mediate discussions
- Provide guidance through choice of tools
- Show where we are going next
- Act as the timekeeper
- Establish agenda with the team
- Educate team members
- Review team progress

Recorder

- Take and transcribe meeting minutes
- Distribute transcribed minutes to team members
- Remain a full team member

Team Values

The members of the Strategic Planning Team have identified and agreed upon the following team values through consensus:

- Seek to be inclusive
- No secrecy surrounding team activities
- Open and frequent communication
- No hierarchy: team members are equal and accountable

- Develop consensus on issues
- Mentor others and each other
- Coach

Team Philosophy

Team members have committed themselves to develop a strategic plan through hard work and dedication that will guide the _____ Fire & EMS and align with the visions and goals already established by City Council and the citizens of _____.

Team members have agreed to "Walk the Talk" using a defined set of criteria as a decision filter in evaluating problem areas and transforming those problems into opportunities. The team and the processes used will not be rigid. Instead, they will be dynamic to remain effective in a constantly changing environment. The team will address new challenges and establish consistency and accountability in how the Department delivers customer service through input and participation of departmental personnel. The team will not be involved in daily operational issues but will work to create the future and improve the system.

Team Resources

Meeting Times and Frequency:

Meeting Location:

Supplies/Equipment:

Personnel Resources:

Accountability

The team members realize the importance of this undertaking. They have identified and established team values, guidelines, and boundaries to ensure accountability among themselves and all members of the Department.

We, the Strategic Planning Team members, commit ourselves as students, mentors, and partners to the vision, purpose, and goals of this charter:

Signature Lines

Index

About the Author

Cecil V. "Buddy" Martinette Jr. is owner of PB&T Consulting, and former chief executive officer, and founding member, of Spec Rescue International, a specialized rescue training and consultation company headquartered in Virginia Beach, Virginia.

Prior to this position, Buddy served in the public sector as chief of the Wilmington Fire Department in Wilmington, North Carolina (retired); assistant county administrator in Hanover County, Virginia; and chief of the Lynchburg Fire and EMS Department in Lynchburg, Virginia.

Buddy started his career at age 15 as a volunteer at the Thalia Volunteer Fire Department in Virginia Beach and 4 years later joined Virginia Beach Fire Department, where he spent 25 years as a firefighter, battalion chief, and chief fire marshal.

Buddy is a former instructor with the State of Virginia Department of Fire Programs, Federal Emergency Management Agency (FEMA) Rescue Specialist instructor, and Incident Support Team operations officer and task force leader for Virginia Task Force II of the FEMA Urban Search & Rescue Program. His Urban Search & Rescue experience includes the Colonial Heights Walmart Collapse; deployments for Hurricanes Floyd, Fran, Frances, and Ivan; the Murrah Federal Building Bombing; and the Pentagon terrorist attacks on 9/11.

Buddy was appointed by Governor Pat McCrory in 2016 to serve as a commissioner on the North Carolina State Emergency Response Commission and by the North Carolina General Assembly to serve on the State of North Carolina 911 Board. He is also a past first vice president of the North Carolina Association of Fire Chiefs and past first vice president of the Southeastern Association of Fire Chiefs. Buddy is also the past president of the Wilmington Central Rotary Club.

Buddy has written numerous articles for the fire service and frequently lectures on organizational leadership and specialized rescue operations to public safety, military, industrial, and law enforcement organizations. Buddy is the author of the first, second, and third editions of the Jones & Bartlett book *Trench Rescue: Principles and Practice to NFPA 1006 and 1670*. He was also awarded a design patent for his nut, bolt, and drill bit sizing tool called "Sizing Buddy."

Buddy has a bachelor of science in fire administration from Hampton University and a master's in public administration from Troy State University. In addition, he is a graduate of the National Fire Academy Executive Fire Officer Program, where he received the Outstanding Research Award in leadership. He has also received the designation of Chief Fire Officer by the Commission on Chief Fire Officer Designation.

He can be reached by email at Budelbow@aol.com.